Adhesives - Properties, Modifications, Typical and Innovative Applications

Edited by Anna Rudawska

Published in London, United Kingdom

Adhesives - Properties, Modifications, Typical and Innovative Applications
http://dx.doi.org/10.5772/intechopen.1004983
Edited by Anna Rudawska

Contributors
Anna Rudawska

Asmare Tezera Admase, Brittany McDonough, Bruno Traber, Desalegn Abera Mersha, Ejigayehu Desalegn Asrade, George W. Kay, Heiko Jung, Michael C. Brown, Mohammad Al-Zu'bi, Nicolae Alin Panc, Rahul Jadia, Raphael Schaller, Sourabh Boruah, Tesfa Nega Gesese, Zenamarkos Bantie Sendekie

First published in London, United Kingdom, 2025 by IntechOpen
IntechOpen is the global imprint of INTECHOPEN LIMITED, registered in England and Wales, registration number: 11086078, 167-169 Great Portland Street, London, W1W 5PF, United Kingdom

For EU product safety concerns: IN TECH d.o.o., Prolaz Marije Krucifikse Kozulić 3, 51000 Rijeka, Croatia, info@intechopen.com or visit our website at intechopen.com.

British Library Cataloguing-in-Publication Data
A catalogue record for this book is available from the British Library

Adhesives - Properties, Modifications, Typical and Innovative Applications
Edited by Anna Rudawska
p. cm.

This title is part of the Materials Science Book Series, Volume 5
Topic: Surface Science
Series Editor: Chonghe Li
Topic Editor: Anna Boczkowska

Print ISBN 978-1-83634-440-7
Online ISBN 978-1-83634-439-1
eBook (PDF) ISBN 978-1-83634-441-4
ISSN 3049-8856

IntechOpen Book Series

Materials Science

Volume 5

Aims and Scope of the Series

Materials science has always occupied an extremely high position in the human development process. As we explore the oceans of stars, various industries have put forward more stringent requirements for the performance of materials, forcing us to pay more and more attention to the development of new materials. At the same time, the formation of a data-driven scientific paradigm is dramatically shortening the development cycle of new materials. The huge data generated by synergistically combining theories, high-throughput experiments, high-throughput computation, and artificial intelligence is greatly contributing to our ability to utilize materials science to solve real-world problems. The three topics of this book series - Metals and Nonmetals; Composite Materials; and Surface Science - will address important areas of advancement in materials science. There will be a range of interesting works published under these topics.

Meet the Series Editor

Prof. Chonghe Li received his Ph.D. from the Chinese Academy of Sciences in 1995. From 1995 to 2000, he worked as a researcher at the Shanghai Institute of Metallurgy, Chinese Academy of Sciences, where he also served as director of the research laboratory. In 2000, he was appointed Professor at the Institute of High-Performance Computing in Singapore, where he worked on computation and simulation in materials science until 2004. Since then, he has been a professor at the School of Materials Science and Engineering, Shanghai University, China, as well as the director of the Shanghai Specialty Casting Engineering and Technology Research Center. Prof. Li's research focuses on titanium alloys, titanium-aluminum single crystals, intermetallic compounds, theoretical calculations, alloy design, and special refractory materials. His broad scientific expertise is well recognized by the scientific community around the world. He is a member of the editorial board of the journal Metals. As an author, he has published more than 200 peer-reviewed papers, 2 books, and over 40 patents.

Meet the Volume Editor

Anna Rudawska, Prof. Ph.D., D.Sc., Eng. works in Lublin University of Technology in Poland. She is an author or co-author of 410 scientific publications in Polish, English, Ukrainian and Slovak, including patents and patent applications. She is a member of many national and international organizations, e. g., the Polymer Processing Society and Society for Adhesion and Adhesives, and she is also a reviewer and member of the editorial board of international Journals. Her research focuses on the influence of technological and design factors on the strength of polymers and metal adhesive joints, including the analysis of surface treatment and adhesive properties to increase the strength of adhesive joints. An important research area is the modification of epoxy adhesives and epoxy composites.

Contents

Preface

Material joining techniques play an important role in human functioning. One of the most developed and frequently encountered methods of joining materials is bonding, which is an alternative technological and construction solution for other joining methods. A significant advantage is the possibility of joining elements made of materials characterized by various mechanical, physical and chemical properties. Bonding technology contributes, among others, to designing and producing light-weight yet durable structures with specific properties.

Because bonding technology is used in many fields, adhesives are chosen based on specific properties. One key factor in selecting an adhesive is its chemical base, which defines those properties. In many cases, the properties of the adhesives are modified using chemical and physical methods to achieve the specific properties dedicated to specific applications, both material and operational. This also applies to innovative solutions in the aspect of modifying adhesives through the use of various modifying additives affecting different properties of the adhesives. It should be noted that when modifying adhesives by changing certain properties, it is possible to improve or worsen others simultaneously, although sometimes a synergistic effect is achieved.

The increasingly used innovative solutions in designing biodegradable adhesives regarding environmental protection and sustainable development are also worth mentioning. The use of materials made from natural materials and biodegradable additives promotes the production of environmentally friendly adhesive materials.

The diversity of adhesives (defined by their specific properties), as well as the modification of adhesives allowing for obtaining specific properties, contributes to the increase in the areas of application of the adhesives. The adhesives are used in various typical industrial sectors, strictly engineering, such as aviation, automotive, construction, machinery, electronics, yacht industry, etc., and in the footwear, furniture, printing and packaging industries. Adhesives are also very often used in medicine (for the production of tools, equipment, dressing materials, tissue adhesives, and adhesive materials in dentistry), jewellery, advertising, decorative elements, industrial design, etc.

The book aims to present selected information on the properties, possibilities of modifying adhesives and typical and innovative applications of adhesives.

I would like to sincerely thank Tea Jelaca, the Publishing Process Manager, for her professional and competent support in the editorial preparation of this book. Thanks to her commitment, the publication process ran smoothly and ended successfully.

Anna Rudawska
Faculty of Mechanical Engineering,
Department of Production Computerisation and Robotisation,
Lublin University of Technology,
Lublin, Poland

Section 1

Classification and Properties of Adhesives

Chapter 1

Introductory Chapter: Classification of Adhesives

Anna Rudawska

1. Introduction

Adhesives are chemical substances consisting of atomic groups called mers. Adhesives are characterized by the fact that, thanks to their adhesion and internal cohesion, they permanently connect various materials without damaging their structure [1, 2]. In order to prepare the adhesive, polymers are used that have high adhesion to the materials being joined and appropriate cohesion after solidification or curing [3]. Adhesives should be characterized by low surface tension during wetting of the material surface. Preparation of the surface layer, type of adhesive, or curing conditions are some of the many factors influencing the properties of the adhesive joints [4, 5].

The chemical structure of the basic component of each adhesive (polymer, monomer, or resin) affects the adhesion phenomenon (when joining elements), joint cohesion, and the resistance of the joint to external factors (temperature, water, oxygen, UV radiation, and various chemical factors). The degree of polymer cross-linking in the adhesive joint and its spatial structure are of great importance [6, 7].

In polymer, resin and monomer molecules there are groups with a specific chemical structure. The most common groups or substituents are: ester, hydroxy, amine or imine, carboxyl, acetal, epoxy, isocyanate, urethane, urea, amide, nitrile, halogen. The basic types of polymers, resins and monomers containing the above-mentioned groups or substituents constituting binders of various adhesives include: polyvinyl acetate (PVAC) and vinyl acetate copolymers, polyacrylates and polymethacrylates and some copolymers, polyvinyl acetals, polyamides (PA), polyvinyl chloride (PVC), rubbers, saturated and unsaturated polyester resins, polyurethanes (PUR) and isocyanates, phenolic and amino resins, epoxy resins, acrylic and methacrylic monomers [8, 9].

There are many types of adhesives, with different properties, applications and conditions for making joints. Considering the diversity and lack of universality, there are several criteria for dividing adhesives. The most important of them are [10–12]:

- Origin of the main component,
- Form of the adhesive,
- Number of components forming the composition,

- Type of curing,
- Strength of the joints,
- Scope of application and others.

2. Origin of the main component

Below is a division based on the origin of the main component:

1. Adhesives of natural origin:
   - Plant (dextrin, starch, and rosin-based),
   - Animal (skin and bone),
2. Adhesives of synthetic origin:
   - Based on synthetic resins (e.g., epoxy and polyurethane),
   - Based on cellulose derivatives,
   - Based on natural rubber and synthetic rubbers.

Among synthetic adhesives, the following can be distinguished:

- Amine – for wood,
- Epoxy – for ceramics and metals,
- Phthalic – for paper, wood, etc.,
- Polyacrylic – for ceramics, metals, plastics,
- Polyurethane – for metals, leather, plastics,
- Polyvinyl – for wood, leather, etc.
- Silicone – for connecting metal with rubber or silicone material, they are resistant to temperatures of ~50°C, and special varieties will withstand even ~800°C.

3. Form of the adhesive

The division based on the form of the adhesive includes liquid and solid adhesives (**Figure 1**).

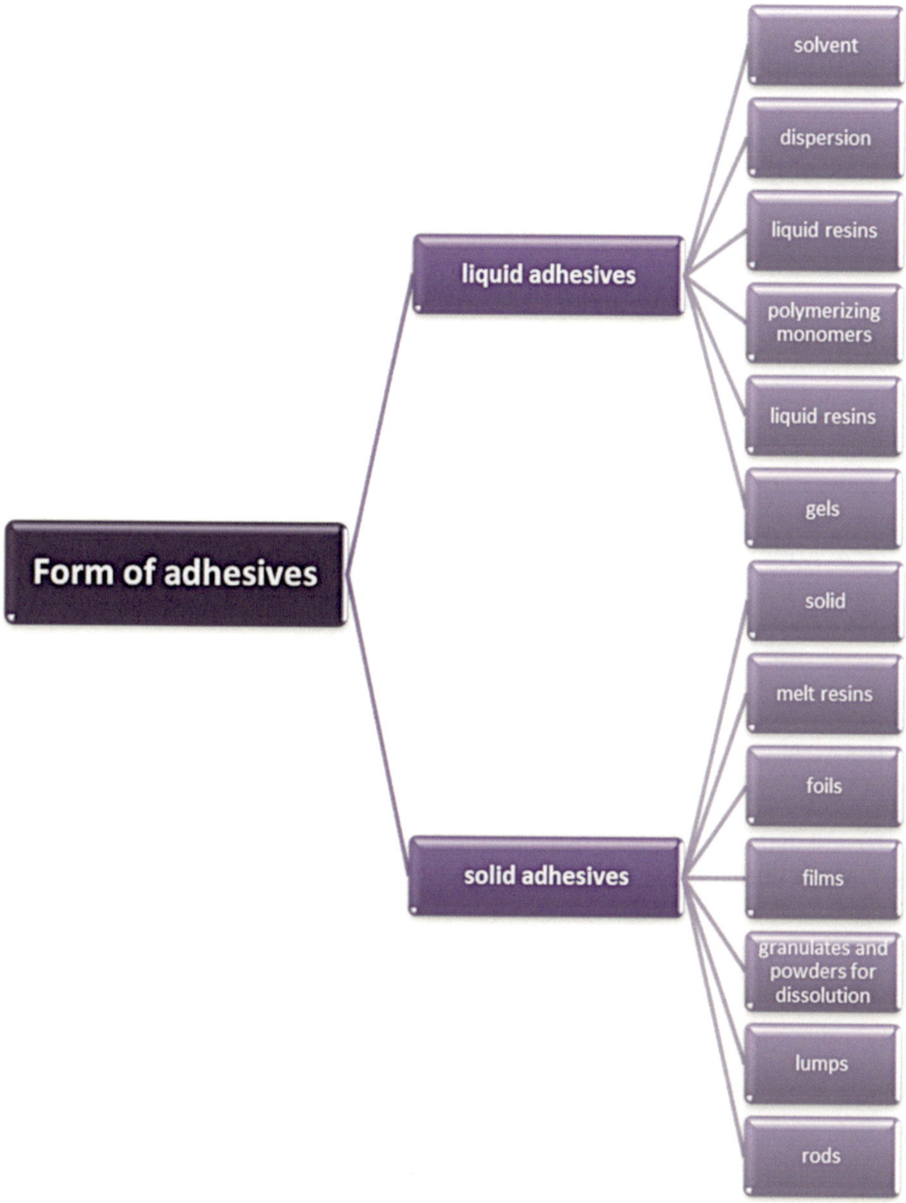

Figure 1.
The division based on the form of the adhesive.

4. Type of curing

The division based on the type of curing is as follows [8]:

1. Adhesives that bind as a result of a *physical process* (e.g., evaporation of the solvent or water). Curing adhesives are characterized by the fact that they penetrate deeply into the material, which causes its swelling and partial dissolution.

After bonding the elements, the surfaces of the bonded materials penetrate each other, and the solvent evaporates, leaving a permanent bond. Curing adhesives include [8, 13, 14]:

a. Solvent adhesives – which are solutions of thermoplastic polymers and are often used when bonding thermoplastics,

b. Dispersion adhesives – two-phase mixtures that contain a suspension of crushed synthetic resin binders, such as: polyacrylate, polyvinyl acetate, or polyurethane. These adhesives are used for bonding wood, paper, and floor coverings.

3. Adhesives that bind as a result of *chemical reactions*:

a. Chemically cured adhesives (curing is the result of chemical reactions),

b. Thermosetting adhesives (curing is the result of high temperature),

c. Anaerobic adhesives (curing occurs rapidly after oxygen is cut off),

d. Cyanoacrylate adhesives (curing occurs under the influence of moisture that remains on the bonded surfaces).

Chemically cured adhesives do not penetrate deeply into the material but are characterized by high strength of adhesive joints. Bonding occurs as a result of a chemical reaction of synthetic curable resins. Chemically cured adhesives react only after the connection is formed, but before this happens, all reactions are blocked by chemical or physical methods. Blocking using physical methods consists in combining the components of the adhesive only before the bonding process itself. Chemical blocking consists of the fact that all reactive components of the adhesive react only under the influence of stimuli, for example, heating or the availability of light or moisture. The advantage of chemically cured adhesives is the fact that the curing time can be regulated by changing the curing temperature or changing the adhesive compositions. Among chemically curable adhesives, the following types of adhesives can be distinguished [13, 15]:

- Polycondensation adhesives – characterized by the fact that during their curing, low-molecular-weight products are released, for example, carbon dioxide,

- Silicones – these are single-component adhesives that cure at room temperature in the presence of moisture,

- Polymerization adhesives – their monomers combine into polymer molecules during curing, without releasing low-molecular-weight substances. This subgroup of adhesives includes: adhesives for sealing threads (anaerobic adhesives) and adhesives for small surfaces such as plastics, rubbers, or metals (cyanoacrylate adhesives),

- Polyaddition adhesives – curing occurs without releasing low-molecular-weight products. As a result of combining the resin and the curing agent, these adhesives polymerize. Polyaddition adhesives include epoxy and polyurethane adhesives.

5. Strength of the joints

The adhesives can be divided into [14, 16]:

- Adhesive tape – the strength of which is sufficient for reversible connection,
- Assembly adhesive – where the strength allows for a permanent connection of two elements but does not allow for treating this connection as a structural element,
- Structural adhesive – having sufficient strength to consider this connection as a structural element.

6. Physical properties

From the point of view of physical properties, adhesives can be rigid, flexible, non-conductive, or conductive to electric current, heat, electric or magnetic fields, waterproof, resistant or not to aggressive chemical environments, etc.

7. Scope of application

Depending on the scope of application, adhesives can be divided into, among others, adhesives for [4, 13, 14, 17, 18]:

1. Technical applications:
    - Bonding metals and metal alloys,
    - Polymer materials,
    - Wood and wood veneering,
- Wall coverings and floor coverings,
    - Wallpapers,
    - Footwear,
    - Regenerative masses, etc.
2. Medical applications:
    - Dental adhesives,
    - Tissue adhesives medical materials (such as plasters),
    - Bonding surgical and other instruments,

- Joining drip equipment,
- Bonding inhalers, etc.

3. Utility and decorative applications:

- Making jewelry: bonding gemstones,
- Bonding amber,
- Making glass tableware with decorative elements,
- Making billboards,
- Bonding road signs,
- Bonding mirrors,
- Bonding packings,
- Bonding toys, etc.

Examples of adhesive and bonding applications in the technical area are given in **Figure 2**.

Adhesives are used in various typical industrial sectors, such as automotive, aviation, construction, machine building, electronics, shipbuilding and yachting, as well as in the footwear, furniture, printing and packaging industries, or toy production and advertising. Adhesives are also very often used in medicine, jewelry, industrial design, etc.

Due to the specific requirements for adhesives for specific applications and conditions of use, there is such a large variety of chemical, physical, technological, and functional properties of adhesives that the right adhesive can be matched to specific requirements.

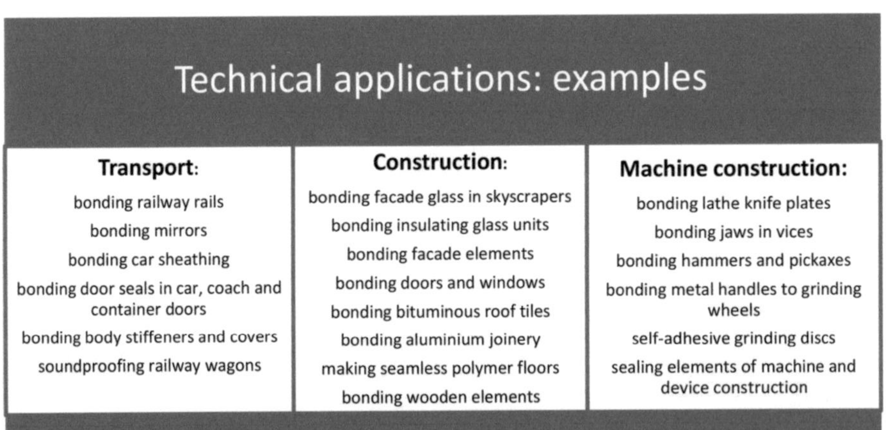

Figure 2.
Examples of adhesive and bonding applications in the technical area.

Author details

Anna Rudawska
Department of Production Computerisation and Robotisation, Faculty of Mechanical Engineering, Lublin University of Technology, Lublin, Poland

*Address all correspondence to: a.rudawska@pollub.pl

References

[1] Wake WC. Adhesion and the Formation of Adhesives. second ed. London: Applied Publishers; 1982

[2] Shields J. Adhesive Handbook. third ed. London: Butterworths & Co (Publishers) Ltd; 1984

[3] Rudawska A, Czarnota M. Selected aspects of epoxy adhesive compositions curing process. Journal Adhesion Science and Technology. 2013;**27**:1933-1950

[4] Adams RD, Comyn J, Wake WC. Structural Adhesive Joints in Engineering. second ed. London: Chapman & Hall; 1997

[5] Rudawska A. Surface Treatment in Bonding Technology. United Kingdom: Academic Press, Elsevier; 2019

[6] Skeist I. Handbook of Adhesives. third ed. New York: Chapmann & Hall; 1990

[7] Hartshorn SR, editor. Structural Adhesives, Chemistry and Technology. New York and London: Plenum Press; 1986

[8] Kinloch AJ. Adhesion and Adhesives. New York: Chapman and Hall; 1987

[9] Pocius AV. Adhesion and Adhesives Technology: An Introduction. Munich: Carl Hanser; 2002

[10] Petrie EM. Handbook of Adhesives and Sealants. New York: McGraw Hill; 1999

[11] Ebnesajjad S. Adhesives Technology Handbook. 2nd ed. New York: Inc. Norwich, William Andrew; 2008

[12] Rudawska A, editor. Adhesives, Applications and Properties. London, UK, Rijeka: Intech; 2016

[13] Adams RD. Adhesive Bonding. Science, Technology and Applications. UK: Woodhead Publishing Limited; 2010

[14] Brockmann W, Geiß PL, Klingen J, Schröder B. Adhesive Bonding. Materials, Applications and Technology. Weinheim: Wiley-Vch Press; 2009

[15] Lee L-H, editor. Adhesive Bonding. 1st ed. New York: Springer Science+Business Media; 1991

[16] Ebnesajjad S, Landrock AH. Adhesives Technology Handbook. London: William Andrew; 1985

[17] da Silva LFM, Öchsner A, Adams RD. Handbook of Adhesion Technology. First ed. Heidelberg, Berlin: Springer; 2011

[18] Mittal KL, editor. Progress in Adhesion and Adhesives. Vol. 5. NY, USA: Scrivener Publishing LLC, Wiley; 2020

Section 2

Modifications of Adhesives

Chapter 2

Advancements in Biomass-Derived Adhesives: Enhancing Overall Properties for Versatile Applications

Asmare Tezera Admase, Zenamarkos Bantie Sendekie, Tesfa Nega Gesese, Desalegn Abera Mersha and Ejigayehu Desalegn Asrade

Abstract

The increasing demand for sustainable materials has propelled research into biomass-derived adhesives as eco-friendly alternatives to traditional synthetic options. This chapter investigates the enhancement of physicochemical and thermomechanical properties of bio-based adhesives, highlighting their applications across diverse industries, including medical devices, construction, automotive, wood and wood-based panels, and packaging. Bio-adhesives, sourced from renewable materials such as lignin, tannins, and proteins, offer advantages like biodegradability and non-toxicity. However, challenges such as low water resistance, bonding strength, and thermal stability persist. Recent advancements, including the incorporation of nanofillers and cross-linking agents, have shown promise in overcoming these limitations, significantly improving adhesive performance. In medical applications, bio-adhesives must be biocompatible and biodegradable to facilitate wound healing without harmful effects. In construction and automotive sectors, they provide sustainable bonding solutions for lightweight materials, enhancing structural integrity under various environmental conditions. The packaging industry is also benefiting from bio-adhesives derived from proteins like whey and casein, which improve barrier properties while remaining renewable. Despite their potential, the commercialization of biomass-derived adhesives faces challenges such as higher production costs and variable performance. Ongoing research aims to address these issues through physicochemical modifications and the development of standardized testing methods. This chapter underscores the transformative potential of bio-based adhesives, emphasizing their role in promoting sustainability while meeting the rigorous demands of various applications. Ultimately, enhancing the properties of these adhesives can contribute significantly to a more sustainable future in material science.

Keywords: biomass-derived adhesives, sustainability, physicochemical properties, thermomechanical properties, eco-friendly materials

1. Introduction

The increasing demand for wood products has prompted a shift toward sustainable forest management. In many regions with limited forest resources, there is a focus on utilizing underused wood and non-wood materials for value-added products [1, 2]. The wood composite industry is also investigating waste and lesser-known species as raw materials. Urea formaldehyde (UF), commonly used as an adhesive, raises health and environmental concerns due to its formaldehyde emissions. An alternative is starch, found in plants like potatoes and rice, which varies in chemical structure based on its amylose and amylopectin ratios. Wood composites such as plywood and particleboard are vital for furniture and construction, traditionally using synthetic adhesives like UF. However, rising environmental concerns and regulatory pressures are driving interest in biomass-based adhesives [3, 4].

Adhesives have been utilized for thousands of years, primarily derived from natural materials like bones, skins, fish, milk, and plants until around 100 years ago. Since the early 1900s, synthetic polymer-based adhesives have emerged, leading to a wide range of industrial applications. Adhesives (bonding agents) are substances capable of attaching to various substrates, enabling the joining of different materials. They are widely used for bonding materials such as wood, metals, and plastics, with applications spanning sectors like automotive, construction, electrical and thermal conductivity, and dentistry. Primarily composed of polymers and various additives, adhesives often include fillers solid, insoluble substances that enhance their performance [5, 6].

Fillers serve multiple purposes, such as reducing costs, modifying mechanical properties, increasing viscosity, and improving electrical and thermal conductivity, as well as enhancing adhesion [7–9]. Common types of fillers include bio-based options like cellulose nanoparticles and tree bark powders; carbon-based materials such as carbon black, carbon nanotubes, diamond, and graphene; ceramics like aluminum oxide, boron nitride, iron carbide, and silicon carbide; and metallic fillers including aluminum, aluminum oxide, copper, silver, and titanium dioxide. The type and quantity of filler added significantly influence the properties of the adhesive, making careful selection essential for optimal performance. Fillers are solid, relatively nonadhesive substances that remain insoluble in the adhesive binder. Their incorporation can modify the rheology and various properties of the adhesive [10–12]. Fillers provide numerous advantages, such as reducing formulation costs, creating adhesives with the desired consistency, and enhancing overall strength. They also contribute to the body of liquid adhesives, improving workability and preventing excessive penetration into substrates. Furthermore, fillers can increase thermal and electrical conductivity, enhance heat resistance, improve solvent resistance, and reduce moisture absorption, among other properties, tailored to meet the specific requirements of various applications [10, 13, 14].

In this decade, plant-derived adhesives prepared from biomasses such as soybean meal (SM) and tannin have been commercially applied in wood composites. Further, SM-based adhesives are considered to have the most potential for replacing petroleum-based adhesives in future because they are abundant, environmentally safe, renewable, and degradable. Many methods have been employed to enhance the bonding performance of plant-based adhesives, including mainly chemical denaturation and modification in the presence of alkalis, surfactants, or organic solvents, cross-linking by cross-linkers or synthetic resins, functional group modification, enzymatic treatment, thermal treatment, ultrasound treatment, and combinations of these methods.

Fillers are materials added to polymer systems to reduce costs or enhance their properties. They can exist in solid, liquid, or gas forms. For instance, a small percentage of clay can significantly improve mechanical and thermal properties [4, 15, 16]. Fillers are generally categorized into two types: inert fillers and active fillers. Inert fillers, derived from inorganic mineral powders like kaolin, diammonium phosphate, porcelain clay, sodium silicate, and magnesium oxide, primarily serve to lower costs. These hydrophilic materials can be effectively dispersed in adhesives. Active fillers, on the other hand, are organic compounds that swell when dissolved in a solution, contributing additional properties to the polymer system [16–18]. According to the findings of reference [19], the addition of copper in amounts of 3–5 wt% significantly improved the bond strength between treated substrates, achieving more than a threefold increase compared to the original adhesive. Furthermore, another strategy to enhance the properties of adhesives based on nitrile butadiene rubbers (NBRs) involved employing a polymer matrix that combined NBR with thermoplastics [4, 19]. All laminating adhesives should have high tack, low substrate penetration, and noncurling properties. Dextrin adhesives remain popular for their cost-effectiveness, while plant-based starch-derived adhesives are commonly used as binders and sizing agents in the paper and textile industries [13]. A common method for reinforcing and toughening epoxy resin is to introduce a suitable filler. Some researchers examined the effects of adding mineral fillers on the mechanical properties of epoxy composites, demonstrates that, it improves the thermal stability of the composite [20, 21].

Polymer nanocomposites are commonly used as adhesives across various fields, including automobile engineering, construction, electronics, and aerospace. Their unique properties significantly enhance adhesion and performance, making them ideal for demanding applications in these industries. To further improve application performance, reinforcing additives such as graphene oxide, silica, carbon black, fibers, and other inorganic metal oxides, ranging from nano to micron sizes, have been incorporated into epoxy resin [22, 23].

Polymeric coatings are commonly used to protect substrates from chemical, physical, and biochemical degradation. The most studied base polymers for these applications include epoxies, polyurethanes, acrylics, and silicones, known for their excellent processability, chemical resistance, and strong adhesion. However, their long-term durability can be limited. Incorporating nanofillers effectively enhances the performance of these polymers, improving mechanical properties such as crack resistance, wear resistance, and corrosion resistance [8, 24–26].

The primary advantages of incorporating additives into adhesives include enhancing bonding strength and reducing costs. The push to lower expenses, combined with the demand for environmentally friendly and health-conscious products, has led to the recycling of waste materials and the use of more affordable natural materials as additives. This trend not only improves adhesive performance but also promotes sustainability in adhesive formulations [16, 27]. The incorporation of fillers in rubber-based adhesives can enhance adhesive strength through a viscoelastic energy loss mechanism, along with potentially other secondary mechanisms. This improvement contributes to the overall performance and durability of the adhesive, making it more effective in various applications [28, 29].

In this decade, plant-derived adhesives (As shown in **Figure 1**) made from biomasses are particularly promising as a replacement for petroleum-based options due to their abundance, environmental safety, renewability, and degradability. Various methods have been employed to enhance the bonding performance of adhesives,

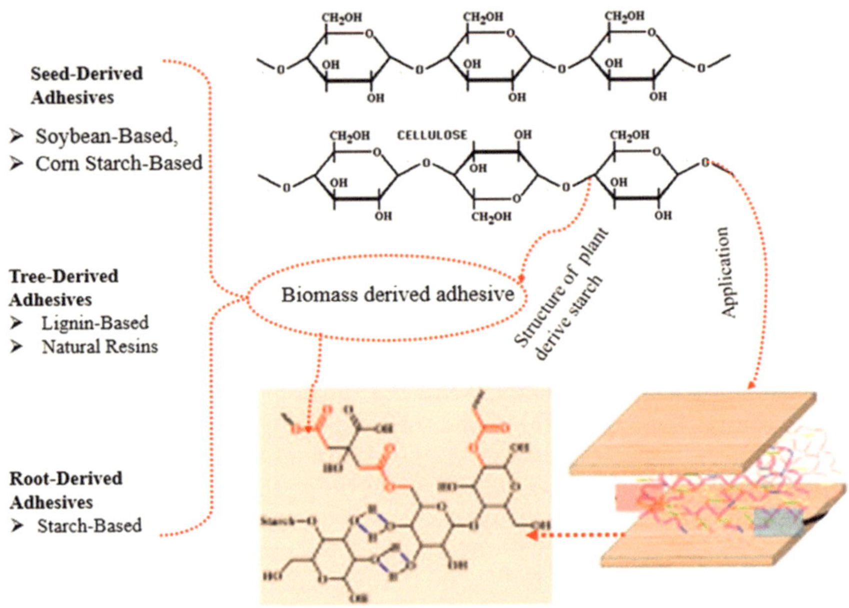

Figure 1.
Structure of plant derived starch based adhesive and their application [30].

including chemical denaturation with alkalis, surfactants, or organic solvents; cross-linking with cross-linkers or synthetic resins; functional group modification; enzymatic treatment; thermal treatment; and ultrasound treatment. Many of these techniques can be combined for optimal results. Fillers with diverse compositions and structures can effectively reduce costs while enhancing the properties of modified polymeric matrices. These improvements can include better processability, reduced density, altered thermal expansion and conductivity, enhanced flame resistance, changes in optical properties, and improved electrical, magnetic, and mechanical characteristics. Notably, nanoscale fillers can significantly boost the performance of adhesives.

2. Physicochemical properties of biomass-derived adhesive

In industries like furniture manufacturing, building, electronics, and packaging, adhesives play a crucial role. Adhesives are a crucial component of plywood and have a direct impact on its functionality and quality [31]. Consequently, much research has been done to enhance the moisture-repellent properties of adhesives, and the cross-linking modification has shown a high enhancement in performance [32]. Bio-based adhesives have some drawbacks, such as low moisture tolerance, low bio-degradation resistance, and low gluing strength. These shortcomings can be alleviated by incorporating renewable materials as fillers and applying different types of cross-linkers into bio-based adhesives [33], chemical modification of nano-cellulose-based wood adhesives, silane modification of starch-based wood adhesives [34], structural modification through physical, biological and chemical methods [35], modification and functionalization of starch [36], and blending with other polymers [37]. Nanofillers, such as clay [38], sepiolite, montmorillonite [39], graphene oxide [40], carbon nanotubes [41], and nanocellulose, have proven to be effective candidates for improving cohesion and bonding strength of adhesive [42]. Enhancing the properties of biomass-derived adhesive offers adverse benefits including fast curing, excellent dry/wet bonding strengths, good tolerance to substrate moisture content [43], improved adhesion properties [26], resistance to aging, heat resistance [44], stable bond performance [45], and improved mechanical properties [46, 47]. Lignin modification either with poly(butylene succinate) or polypropylene biocomposites increased the thermal stability of kenaf core fiber [48].

Due to the chemical cross-linking between epoxy groups and protein molecules, Luo et al. [49] discovered that adding 8% triglycidylamine (TGA) improved the water resistance of soybean meal-based (SM) adhesive by 15.1%. The SM/TGA adhesive's water resistance increased by 24.6% when 8% acrylic emulsion (AE) was added, as a result of an interpenetrating network that was formed. The adhesive's solid content increased to 30.44% with the addition of 8% TGA and to 34.92% with 8% AE. Through restricting the evaporation of water, the increased solid content decreased the cohesive failure of the cured glue.

An eco-friendly adhesive based on soybean meal that is improved with diglycidyl resorcinol ether (DRE) is investigated by Luo et al. [32]. The study uses FTIR measurement to describe the chemical reaction between DRE and soybean meal (SM) adhesive. The neat soybean meal adhesive's FTIR spectra revealed that the primary absorption bands were roughly located at 1635 cm^{-1} (amide I), 1523 cm^{-1} (amide II), and 1240 cm^{-1} (amide III) (**Figure 2**). When DRE was added, amide I, amide II, and amide III shifted toward a higher wavenumber, roughly at 1652, 1535, and 1265 cm^{-1}, respectively. This suggests that the hybrid adhesive formed a more stable structure when compared to the neat soybean meal adhesive because of the interaction and reaction between DRE and soy protein. DRE addition to adhesives caused a ring-open reaction in the epoxy group, with peaks at 908 cm^{-1} not visible. A new peak formed around 1741 cm^{-1} after addition. The peak at 2922 and 2856 cm^{-1} corresponds to the carbonyl group of the ester bond, resulting from the esterification process between DRE and soy protein. These modifications revealed that, during hot pressing, DRE interacted with the functional groups of the soybean meal adhesive to create a cross-linking network in a ring-opening process *via* a covalent connection, as shown in **Figure 2**. A more even and smooth surface was achieved in the soybean meal-based adhesive by the incorporation of DRE, which has a flexible chain structure and benzene ring. This effective cross-linker created covalent networks with soy protein molecules (**Figure 2**).

Chen et al. [50] studied soybean meal-based wood adhesives enhanced by phenol hydroxymethylated tannin oligomer (PHTO) for exterior use. The study found that the residual rate of cured adhesives rose and fell with PHTO addition. Using soybean meal-based adhesives (SPAs) containing 10% PHTO, the residual rate increased to 89.2%, 11.6% higher than pure soybean meal. The high solid content in PHTO caused the SPAs solid content to grow from 27.6% to 29.6%. The moisture absorption rate of SPAs decreased with PHTO addition, but too much PHTO allowed the system to absorb more moisture.

Chen et al. [51] used a larch tannin-based resin (TR) to improve the performance of adhesives based on soy protein. The outcome demonstrates that adding 5 weight percent TR to soy protein-based adhesives (SPAs) boosted the adhesive's residual rate by 38.4% and the plywood's wet shear strength by 105.4%. The amount of unfolded protein molecular chains in the resulting adhesive, which gave it an extremely high viscosity, reduced the bonding strength of the adhesive when 10% weight percent

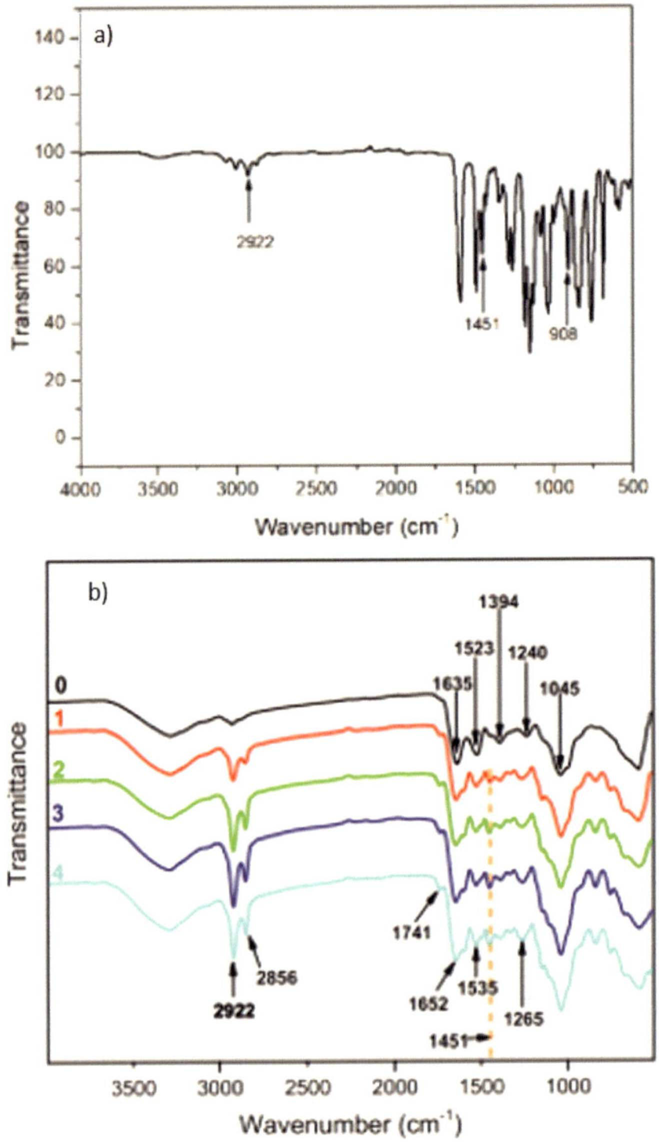

Figure 2.
(a) FTIR spectra of DRE. (b) FTIR spectra of the different adhesives: 0 (SM adhesive), 1 (SM/DRE-2 adhesive), 2 (SM/DRE-4 adhesive), 3 (SM/DRE-6 adhesive), 4 (SM/DRE-8 adhesive) (taken from Ref [32]).

TR was added to (SPAs). As a result, the glue did not adhere to the wood surface consistently, which prevented the plywood from having an effective bond. The mechanisms behind the improved water resistance and cross-linking density of the adhesive are as follows: (a) a cross-linking structure formed by the reaction between the soy protein molecules and TR; (b) an interpenetrated network formed by the cross-linking protein molecules and the self-cross-linking TR molecules; and (c) NaOH in TR increased the soy protein's solubility and ionic amino acid content, increasing the soy protein's reactive sites with cross-linkers and, consequently, the cross-linking density.

Yuan et al. [52] constructed high-performance wood adhesive from functionalization of cellulose with amine group cross-linked with branched epoxy. The outcome showed that a dense polymer cross-linking network formed by covalent and secondary bonds between molecules lessens the infiltration of water molecules.

According to Zhu et al. [53], adding tannic acid (TA) as a cross-linker to soy protein glue increases the adhesive's water resistance. The results showed that the wet-resistance of adhesives was strongly impacted by the phenolic hydroxyl (catechol) level in the SP-TA solution was also noteworthy. The hydroxyl group in phenolic compounds is limited to the tyrosine residue in virgin SP, with a hydroxyl concentration of 62.9 GAE/g. The free phenolic hydroxyl content for SPT (240.7 GAE/g) rose as predicted upon the addition of TA. A crucial rheological characteristic that affects protein dispersions' wetting and spreading behaviors and, in turn, their adhesion performance is viscosity. In comparison to SP and SPT, the viscosities of SPT-Zn, SPT-Fe, and SPT-GH were greater; SPT-Fe had the highest value, 2397.6 cp. The reduction of the viscosities of SPT-Zn and SPT-GH to 921.5 and 1120.6 cp, respectively, suggests that the coordination bond had a stronger effect on the shearing behaviors of SPT complexes than the hydrogen and covalent bonds did.

Zidane et al. [54] explore the preparation and characterization of tannin-based adhesives reinforced with cellulose nano-fibrils (CNFs) for wood bonding. The strong reactivity of tannins with the cross-linking agent is demonstrated by gel time measurements. The glue control and the sample containing 1% CNFs were not statistically similar to the adhesives containing 5 and 10% CNFs. The pH values, a crucial element that controls the adhesive hardening during bond process pressing, revealed no statistically significant difference between the control and the 10% adhesives. The outcome demonstrates that viscosity increases with increasing CNF content. However, the viscosity exhibited a substantial rise with the 5 and 10% CNFs adhesives. The distinction in the molecular weights of cellulose and tannin can help to explain this. The viscosity and solids content of the adhesives were mostly impacted by the addition of CNFs.

The broadband peak in the research by [55] moved from around 3276 to 3367 cm^{-1} upon the addition of TCA-HBSi, indicating that hydrogen-bonding initiated cross-linking structure was formed in SP/TCA-HBSi resins. Furthermore, with the addition of TCA-HBSi, the composite resin's C−C/C−H percentage rose from 54% to 68%. This peak denotes the effective increase of intramolecular contacts in cured resin and correlates to the hydrophobic qualities and degree of cross-linking of the protein structure.

Pradyawong et al. [56] found that the addition of lignin increased the spreadability and thermal stability of soy protein adhesives. Lignin is thought to have significant promise for enhancing the water resistance of adhesives based on proteins. Hardwood kraft lignin phenolation resulted in a formaldehyde emission that was well below the crucial threshold [57]. The investigation also showed that lignin's chemical structure remained alert during phenolation. Following phenolation, there was an increase in the peak at 1222 cm^{-1}, which was attributed to the newly added phenol groups' phenolic C−O stretching vibration.

Jin et al. [58] applied multiple cross-linking strategies to achieve high bonding strength and antibacterial properties of double-network soy adhesive. In contrast to SM/HBSi, which had an insoluble percentage of 73.62 ± 2.11%, SM adhesive had an insoluble fraction of 69.45 ± 2.32%. Furthermore, a considerable drop in the swelling ratio was seen between the SM/hyper-branched silicone (HBSi) adhesive and that of SM, indicating a noticeable improvement in water resistance. Due to the SieOeSi moieties in the integrated HBSi, which formed a highly water-resistant area and prevented water entry, this improvement was possible. As the HBSi concentration grew, the insoluble percentage of the SM/TCu/HBSi adhesives also increased, and the swelling ratio for SM dropped from 413.8 ± 25.7% to 52.4 ± 19.2%. These alterations are mostly associated with the cross-linking density and hydrophilic group content.

Sain et al. [59] investigate the effect of curing temperature on the properties of lignin and tannin based bio adhesive. The findings demonstrate that greater temperature curing ensured the higher cross-linking between the resin components. As a result, the resin that was cured at 200°C demonstrated improved thermal stability and a reduced curing time (30 minutes).

Chang et al. [60] added fillers, such as pulp fiber (PF), bagasse fiber (BF), and poplar wood fiber (WF), to the adhesive based on soybean meal to improve its performance by creating a hydrogen connection between the PF and the soybean meal system. It was discovered that the residual rate dropped from 74.70% to 70.74% when WF was added to the CSM glue. BF was added to the glue, causing the residual rate to drop from 74.70% to 70.45%. The residual rate of the CSM adhesive, on the other hand, increased from 74.70% to 82.91% when 0–4% weight percent PF was added.

Through multi-interfacial interactions, the right PF addition can efficiently create a cross-linking structure with other molecules in the adhesive, increasing the glue's water resistance. Following the addition of WF and BF, it was noticed that the glue's viscosity dropped. This may be attributed to the ineffective binding of WF and BF with the adhesive molecules. Following the addition of 0–4% weight percent PF, the adhesive's viscosity rose. This response demonstrated that the PF reinforced the cross-linking network through repeated contacts with the sticky molecules.

Using xylan and tannic acid produced from biomass and cross-linked with acrylic acid to impart antioxidant and antibacterial qualities, Chang et al. [61] created sticky hydrogels inspired by mussels. As a consequence, the hydrogel was shown to exhibit high biocompatibility along with strong antioxidant and antibacterial properties.

3. Thermomechanical properties of bio-based adhesives

The study of the thermomechanical characteristics of bio-based adhesives is an emerging area of research that corresponds with the growing need for eco-friendly materials across industries, especially in the building and packaging sectors. Recent innovations in the development of bio-based adhesive compositions have shown notable enhancements in bonding strength, resistance to heat, and overall mechanical efficacy [62–64]. For instance, adhesives formulated from glycyrrhizin acid and soybean oil have shown comparable properties to traditional adhesives, highlighting their potential for self-healing and recyclability [65]. Additionally, the incorporation of lignin and tannins into adhesive formulations has been shown to enhance thermal properties and water resistance, making these materials suitable for demanding applications [66–71]. The thermomechanical properties of bio-based adhesives are crucial for their performance in various applications.

3.1 Thermal properties

Improving the thermal characteristics of adhesives derived from biodegradable sources is a crucial research focus, especially as industries look for eco-friendly substitutes to replace products made from petroleum [62, 66, 72, 73]. One promising approach involves the incorporation of lignin into soy protein adhesives. Pradyawong et al. [56] highlighted that adding kraft lignin, known for its hydrophobic properties and high reactivity, significantly enhances soy protein adhesives' water resistance and adhesion properties. This modification improves performance and reduces costs due to lignin's abundance and low price [56, 70, 74], significantly enhancing soy protein adhesives' water resistance and adhesion properties. Similarly, Jalali [75] reported that bio-based polymers synthesized from liquefied biomass exhibited comparable or superior performance to petroleum-based adhesives in terms of bond strength, thermal stability, and water resistance, indicating that these materials can meet or exceed industry standards.

Another innovative method is the development of bio-based epoxy resins from agricultural waste, such as banana fibers. Li et al. [72] demonstrated that these resins exhibited shear strengths significantly higher than conventional epoxy systems, suggesting a broader application potential for bio-based adhesives in various structural contexts. Furthermore, the use of bio-based polyols, such as 1,5-pentanediol, has been shown to yield polyester polyols with thermal properties comparable to their petroleum-derived counterparts, thus enhancing the performance of bio-based adhesives [64, 76–79]. Moreover, the thermal properties of bio-based polyurethane adhesives synthesized from isocyanate-functionalized cellulose acetate and castor oil have been characterized using differential scanning calorimetry (DSC) and thermogravimetric analysis (TGA) [71, 80–82]. These analyses indicated that the bio-based formulations exhibited favorable thermal stability, making them suitable for wood bonding applications [83]. The thermal conductivity of insulation boards made with bio-based adhesives has also been evaluated, revealing that these materials can achieve low thermal conductivity values, thus enhancing their insulation properties [81, 84]. In addition, other bio-based materials such as epoxidized soybean oil have been explored for their adhesive properties. Results have indicated that these adhesives possess flame retardance and thermal stability comparable to conventional pressure-sensitive adhesives (PSAs) [85, 86]. The dynamic mechanical analysis of these adhesives revealed a robust thermal response, which is essential for applications requiring durability under varying temperature conditions. A starch-based adhesive that was reinforced with kaolin clay showed a better mass loss in TGA analysis [30] and comparable performance with other properties of adhesives.

In summary, the thermal properties of bio-based adhesives are influenced by their composition, formulation, and processing conditions. Ongoing research continues to explore various bio-based materials, such as lignin, soy protein, and epoxidized oils, to enhance the performance of these adhesives, making them viable alternatives to traditional synthetic adhesives in various industrial applications.

3.2 Mechanical properties

Improving the mechanical characteristics of bio-adhesives is a crucial research focus, especially for biomedical applications, such as wound healing and tissue repair. Researchers have investigated various methods to enhance the mechanical performance of these materials, such as chemical alterations [65, 79], cross-linking [72, 84, 87], and the addition of nanomaterials [88]. A particularly effective technique for boosting the mechanical properties of bio-adhesives involves creating hydrogels with double-cross-linked networks. Wang et al. [89] showed that these hydrogels display enhanced tissue adhesion and mechanical performance owing to the combined effects of adhesive and cohesive forces, which are essential for maintaining structural integrity under stress. This dual-network structure results in a more resilient internal framework capable of resisting external forces, thus improving the overall adhesive effectiveness [84, 90]. Furthermore, the use of biocompatible substances, such as polysaccharides, can greatly enhance the mechanical characteristics of bio-adhesives. According to Yang and Wang [91], chemically modifying polysaccharides through various cross-linking processes can result in notable improvements in mechanical strength. This affirmation is reinforced by Bashir et al.'s [92] investigation, which examined the adhesive performance of bio-adhesives derived from protein-polysaccharide phase separation, showing that the mechanical properties can be adjusted to suit specific applications.

In addition to chemical alterations, the integration of nanoparticles has emerged as a promising technique for strengthening bio-based adhesives. Studies by Tserpes and Tzatzadakis [93] suggest that incorporating multi-walled carbon nanotubes (MWCNTs) can improve the mechanical properties of bio-based adhesives, offering a method for developing more robust and long-lasting adhesive systems. This finding is supported by the study of Adelka [94], which demonstrated that adding nanocatalysts to bio-adhesives could significantly improve the bonding qualities of particleboards produced from agricultural waste.

Moreover, creating bio-adhesives with adjustable mechanical characteristics is crucial for applications that demand specific adhesive properties [16, 80, 83, 95]. Feng et al. [96] investigated the significance of dynamic mechanical attributes in bio-adhesives and proposed that customizable yield stress can avert mechanical breakdown and tissue damage during use. This adaptability is vital to ensure bio-adhesives function effectively under various physiological conditions [94, 97]. Additionally, investigating innovative materials, such as catechol-linked chitosan [89, 98], has demonstrated potential in improving the mechanical and adhesive qualities of bio-adhesives. Zeng and Mo [98] noted that bio-inspired catechol-linked chitosan displays robust adhesion to tissue surfaces, which is essential for effective wound healing. This agrees with Yang et al.'s [99] research, which produced highly elastic and adhesive hydrogel dressings that combine mechanical strength with antibacterial properties, further illustrating the potential of multifunctional bio-adhesives.

A cassava starch-derived bio-adhesive with the incorporation of kaolin and tannin was evaluated for the mechanical properties of shear and tensile strength [25]. As shown in **Figure 3A**, the shear strength of the adhesives increased from 2.7 to 4.9 MPa as the amounts of kaolin and tannin increased from 10–50 w/w% to 15–35 w/w%, respectively. Similarly, the tensile strength also increased from 1.11 to 5.76 MPa as

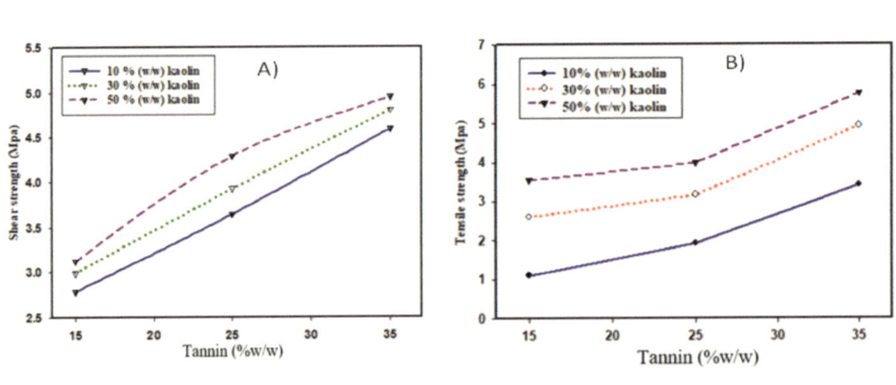

Figure 3.
Effect of kaolin and tannin on the shear strength (A) and tensile strength (B) of adhesive.

the amounts of kaolin and tannin increased from 10–50 w/w% to 15–35 w/w%, respectively (**Figure 3B**). Generally, enhancing the mechanical properties of bio-adhesives involves a comprehensive strategy that encompasses chemical modifications, nanoparticle reinforcement, and the development of adjustable materials. These advancements are essential for enhancing the performance and applicability of bio-adhesives in medical and industrial settings.

4. Applications

4.1 Medical devices

The most popular use of bio-adhesives is wound healing. Despite the possibility of the formation of scar tissue and harm, sutures or staples are traditionally used. As a result, adhesives have been employed as a noninvasive technique to join the wounds of two sides, the tissue beneath the surface, or a combination of the two. The adhesives applied for this application must be properly selected to be biocompatible and biodegradable. The glue's ability to gradually deteriorate over time is crucial in this particular application so that the wound will heal and the adhesive will disappear [100]. Bio-based polysaccharides, such as plasters, are also frequently employed in medical applications. In this instance, their non-toxicity and ability to absorb moisture without losing adherence provide them an advantage over synthetic adhesives [101].

4.2 Construction industry

In a building, adhesives can also be used to join wood beams or panels. The environment is in contact with construction materials. For this reason, research on wood/adhesive combinations especially those that can survive harsh weather conditions has been ongoing to ensure the structure's longevity and resilience [100]. Numerous bio-adhesives have been developed specifically for use in structural applications on hardwood substrates [102]. The demand for more high-throughput products in the construction industry, particularly in the areas of shock, heat, moisture, and UV resistance, has been highlighted in the report "North American and European Construction Adhesives and Sealants Market, Forecast to 2022" [101].

4.3 Automotive industry

Modern lightweight automobiles are designed to have a less carbon impact. Therefore, it is imperative to develop adhesive solutions that bind lightweight materials together as well as adhesives that are low in hazardous compounds and easy to recycle [101].

4.4 Wood and wood-based panel industry/furniture

Tannins, when widely available, offer a promising and economically feasible alternative to synthetic adhesives. Industrial particleboard has been made using tannic resin adhesives in numerous nations. For instance, in South Africa and South America, mimosa tannins have been utilized in the industrial manufacturing of particleboard. To increase tannin-based adhesives' potential for use in the wood-based panel sector, it is necessary to enhance their water resistance and create efficient cross-linking agents in the future [103–105] The potential for tannin-based bio-adhesives to be used in the wood-based panel sector is limited due to their low water resistance. If this property could be improved upon, further applications could be made of them since boards for outdoor as well as internal use might be developed. Researchers have shown that under acidic conditions, the water resistance of a tannin-based bio-adhesive can be enhanced by cross-linking it with epoxy resin, glyoxal, and furfuryl alcohol [64]. A renewable source of adhesive starch is still used in the production of corrugated cardboard. The hydroxyl groups on the starch and the cellulose in the paper are joined by sodium borate ($Na_2B_4O_7{\cdot}10H_2O$) [101].

4.5 Packaging industry

Research on proteins like wheat gluten, soy protein, casein, and whey protein is increasing due to the demand for renewable and biodegradable packaging materials. These proteins can function as adhesives and enhance barrier properties [106, 107]. Whey protein, a cheese by-product, has been extensively studied for packaging applications, demonstrating suitability for edible films and coatings. While these films have a modest water vapor barrier, they offer good oxygen barrier properties. Additionally, whey protein is widely available from cheese production [108]. Casein is often used for adhering labels on glass bottles due to its strong adhesion and easy removal in warm water [101].

5. Challenges and future perspective

The availability of raw ingredients for natural adhesives is dependent on the season and having low bonding strength renders them unsuitable for use in contemporary production. Furthermore, natural adhesives particularly those derived from biomass generally exhibit low water resistance, which causes synthetic adhesives to gradually replace them [109, 110].

Recent research has revealed the significant potential of bio-adhesives in various industries, though challenges like high viscosity, low bonding strength, and poor water resistance remain. Lignin, with its strong self-adhesive properties, offers a viable alternative to phenolic resin adhesives and can help reduce formaldehyde levels. However, its complex structure and limited active groups restrict its performance. Activation modification can enhance lignin's reactivity by increasing active groups or introducing new ones, allowing for the creation of high-performance adhesives. Biomass-derived adhesives offer environmental benefits and are crucial for sustainable development. However, they need improvements in corrosion resistance and drying rate compared to petrochemical adhesives. Expanding research in this area can create new markets while protecting the environment [78, 111].

A variety of adhesive technologies can use renewable materials that may outperform petroleum-based options. However, higher costs and implementation challenges hinder their commercialization, especially when they are not direct substitutes for existing technologies. Renewable resources like starch, casein, and polysaccharides serve as bases for bio-adhesives, offering a foundation for biodegradable adhesives. However, their natural origins affect their structure and performance, requiring modifications for specific applications, particularly regarding water resistance. Methods to enhance water resistance include combining proteins with synthetic adhesives, altering structures for better cross-linking, and using chemical cross-linking for denser networks. Additionally, when biodegradability is not essential, these adhesives are prone to hydrolytic degradation, which can be mitigated by adding hydrophobic groups and increasing cross-linking density through reactions with aldehydes or isocyanates. Bio-based adhesives are competitive with petroleum-based options, though their market is still small and performance reports can vary. They offer a strong alternative to traditional products, with potential for growth and market expansion [1, 101].

Bio-based adhesives are often susceptible to mildew, have low water resistance, and poor stability due to excess reactive groups in their raw materials. To enhance their performance for large-scale applications, physicochemical modifications are crucial. Techniques such as co-mixing, ultrasonication, high-pressure treatment, grafting, esterification, and oxidation have been used to overcome these challenges. Direct use of lignin in adhesives is rare, as these adhesives typically do not meet performance standards and have low water resistance. The dark color of lignin-based adhesives also affects their acceptance. Common methods to improve lignin's water resistance include phenolization, chemical treatment, and creating various network architectures [2, 112].

The product life of bio-based adhesives is often reduced by microbial contamination from moisture and nutrients in natural polymers. To promote their use, it is crucial to prevent mold growth, typically achieved by adding natural anti-mold or antibacterial agents. Although vegetable oil-derived adhesives are environmentally friendly, their production may be more costly than synthetic alternatives due to the need for specialized equipment and oil refinement. This higher cost can hinder adoption, especially in price-sensitive sectors. Additionally, these adhesives often have lower bonding strength, shorter lifespan, and are less resistant to environmental factors like moisture and temperature, which can affect their performance. More development and optimization are needed to ensure they meet specific application requirements [56, 113].

Conventional adhesives are widely used in sectors like construction, automotive, electronics, and aerospace, demonstrating effectiveness in regulated environments. While vegetable oil-based adhesives are emerging as alternatives, their use is limited. They show promise in sustainability-focused areas such as eco-friendly buildings and packaging. However, their adoption is restricted by the reliance on established adhesive technologies requiring regulatory compliance. To promote sustainable alternatives, standardized testing methods addressing current shortcomings are essential, fostering innovation in renewable resource-based adhesives [114, 115].

6. Conclusion

The exploration of biomass-derived adhesives presents a promising avenue for advancing sustainable materials across various industries. This chapter has highlighted the significant improvements in physicochemical and thermomechanical properties achieved through innovative modifications, such as the incorporation of nanofillers and cross-linking agents. These enhancements are crucial for expanding the applicability of bio-based adhesives in sectors like medical devices, construction, automotive, wood and wood-based panels, and packaging industries. Despite the inherent advantages of bio-adhesives, including biodegradability and non-toxicity, challenges such as low water resistance, bonding strength, and production costs remain barriers to widespread adoption. Continued research is vital to address these issues, focusing on physicochemical modifications and the development of standardized testing protocols to ensure performance consistency. As industries increasingly prioritize sustainability, the demand for eco-friendly alternatives to synthetic adhesives will continue to grow. By leveraging renewable resources and enhancing their properties, biomass-derived adhesives can play a pivotal role in creating environmentally conscious products. The future of adhesive technology lies in the ability to balance performance with sustainability, and bio-based adhesives are well-positioned to meet this challenge. Advancing these sustainable materials will not only contribute to environmental stewardship but also open new markets and applications, fostering innovation in the field of adhesive science.

Author details

Asmare Tezera Admase*, Zenamarkos Bantie Sendekie, Tesfa Nega Gesese, Desalegn Abera Mersha and Ejigayehu Desalegn Asrade
Chemical Engineering Program, Faculty of Chemical and Food Engineering, Bahir Dar Institute of Technology, Bahir Dar University, Bahir Dar, Ethiopia

*Address all correspondence to: asmaretezera@gmail.com

References

[1] Mittal KL. Progress in Adhesion and Adhesives. New Jersey: Wiley; 2023

[2] Zhang Z, Macquarrie DJ, Clark JH, Matharu AS. Chemical modification of starch and the application of expanded starch and its esters in hot melt adhesive. RSC Advances. 2014;**4**:41947-41955. DOI: 10.1039/c4ra08027k

[3] Salleh KM, Hashim R, Sulaiman O, Hiziroglu S, Wan Nadhari WNA, Abd Karim N, et al. Evaluation of properties of starch-based adhesives and particleboard manufactured from them. Journal of Adhesion Science and Technology. 2015;**29**:319-336. DOI: 10.1080/01694243.2014.987362

[4] Chang Z, Sun Z, Wu W, Chen T, Gao Z. Effects of inorganic filler on performance and cost effectiveness of a soybean-based adhesive. Journal of Applied Polymer Science. 2020;**137**:1-10. DOI: 10.1002/app.48892

[5] Zhang H, Liu P, Mai C, Zhang K. Dialdehyde cellulose as a bio-based robust adhesive for wood bonding. ACS Sustainable Chemistry & Engineering. 2019;7:10452-10459. DOI: 10.1021/acssuschemeng.9b00801

[6] Aprilliana N, Hidayat W, Asmara S, Lubis MAR, Bakri S, Hidayati S. Effect of natural rubber latex adhesive content on the physical and mechanical properties of agriboard from cassava stem wastes. In: AIP Conference Proceedings. Bandar Lampung, Indonesia: AIP Publishing; 2022

[7] Nemati Giv A, Ayatollahi MR, Ghaffari SH, da Silva LFM. Effect of reinforcements at different scales on mechanical properties of epoxy adhesives and adhesive joints: A review. The Journal of Adhesion. 2018;**94**:1082-1121. DOI: 10.1080/00218464.2018.1452736

[8] Buckner CA, Lafrenie RM, Dénommée JA, Caswell JM, Want DA, Gan GG, et al. We are IntechOpen, the World's Leading Publisher of Open Access books Built by Scientists, for Scientists TOP 1%. Vol. 11. London, UK: Intech; 2016. p. 13

[9] Chen L, Wang Y, Zia-ud-Din FP, Jin W, Xiong H, Wang Z. Enhancing the performance of starch-based wood adhesive by silane coupling agent (KH570). International Journal of Biological Macromolecules. 2017;**104**:137-144. DOI: 10.1016/j.ijbiomac.2017.05.182

[10] Sanghvi MR, Tambare OH, More AP. Performance of Various Fillers in Adhesives Applications: A Review. Berlin Heidelberg: Springer; 2022

[11] Betancourt NG, Cree DE. Mechanical properties of poly (lactic acid) composites reinforced with $CaCO_3$ eggshell based fillers. MRS Advances. 2017;**2**:2545-2550. DOI: 10.1557/adv.2017.473

[12] Nayak SY, Heckadka SS, Sharma P, Aman I, Dey S. Use of egg shell particulate as fillers in E-glass/epoxy composites. In: Proceeding International Conference on Mechanical Engineering and Industrial Automation; Dubai. 2015. pp. 21-25

[13] Vishnuvarthanan M, Rajeswari N. Additives for enhancing the drying properties of adhesives for corrugated boards. Alexandria Engineering Journal. 2013;**52**:137-140. DOI: 10.1016/j.aej.2012.10.002

[14] Abbasi BI. Advancements in Nano-Based Adhesive Bonding An Overview of Nano-Based Adhesive Bonding Incorporation of Nanofillers in Adhesives. 2024. pp. 1-7

[15] Xu H, Zhang X, Yu Y, Yu Y, Yang Z, Zhu X, et al. Enhancement of mechanical properties of epoxy resin matrix adhesives by high-performance fillers. Journal of Polymer Research. 2023;**30**: 1-10. DOI: 10.1007/s10965-023-03755-x

[16] Anaç N. The Effect of Organic Fillers on the Mechanical Strength of the Joint in the Adhesive Bonding. Basel, Switzerland: MDPI; 2023

[17] Wolf A, Buchman A, Eitan A, Fine T, Nevo Y, Heyman A, et al. Improved adhesives containing CNT/SP1 nano fillers. The Journal of Adhesion. 2012;**88**:435-451. DOI: 10.1080/00218464.2012.660398

[18] Anjana S, DSP T. Development of rice husk and egg shell fillers in polymer composites: A review. International Journal of Innovative Science and Research Technology. 2020;**5**:292-294. DOI: 10.38124/ijisrt20aug241

[19] Alekhin AK, Lyusova LR, Naumova YA, Kotova SV, Monakhova TV. Improving the adhesive properties of adhesives based on a mixture of nitrile butadiene and chlorinated rubbers with metal chelates. Polymer Science, Series D. 2020;**13**:26-30. DOI: 10.1134/S1995421220010025

[20] Long J, Li C, Li Y. Enhancement of mechanical and bond properties of epoxy adhesives modified by SiO_2 nanoparticles with active groups. Polymers (Basel). 2022;**14**:1-18. DOI: 10.3390/polym14102052

[21] Belli R, Kreppel S, Petschelt A, Hornberger H, Boccaccini AR, Lohbauer U. Strengthening of dental adhesives via particle reinforcement. Journal of the Mechanical Behavior of Biomedical Materials. 2014;**37**:100-108. DOI: 10.1016/j.jmbbm.2014.05.007

[22] Anoop V, Sankaraiah S, Jaisankar SN, Chakraborty S, Mary NL. Enhanced mechanical, thermal and adhesion properties of polysilsesquioxane spheres reinforced epoxy nanocomposite adhesives. The Journal of Adhesion. 2021;**97**:1-18. DOI: 10.1080/00218464.2019.1620107

[23] Dompé M, Vahdati M, Van Ligten F, Cedano-Serrano FJ, Hourdet D, Creton C, et al. Enhancement of the adhesive properties by optimizing the water content in PNIPAM-functionalized complex coacervates. ACS Applied Polymer Materials. 2020;**2**:1722-1730. DOI: 10.1021/acsapm.0c00185

[24] Asbury JH. A Study of Starch. USA: Georgia Institute of Technology; 1933

[25] Admase AT, Mersha DA, Kebede AY. Cassava starch-based hot melt adhesive for textile industries. Scientific Reports. 2024;**14**:1-11. DOI: 10.1038/s41598-024-70268-y

[26] Soubam T, Gupta A, Jamari SS. Eco-friendly bio-based adhesive for plywood from natural rubber latex (NRL)-blended isocyanate cross-linked starch. Environmental Science and Pollution Research. 2023;**30**:124610-124618. DOI: 10.1007/s11356-022-20788-9

[27] Iijima M, Muguruma T, Brantley WA, Yuasa T, Uechi J, Mizoguchi I. Effect of mechanical properties of fillers on the grindability of composite resin adhesives. American Journal of Orthodontics and Dentofacial Orthopedics. 2010;**138**:420-426. DOI: 10.1016/j.ajodo.2008.08.039

[28] Bhowmick AK. Effect of model fillers on the adhesive strength of pressure-sensitive tapes. Journal of Adhesion Science and Technology. 1989;**3**:371-381. DOI: 10.1163/156856189X00272

[29] Jung H, Kim MK, Lee JY, Choi SW, Kim J. Adhesive hydrogel patch with enhanced strength and adhesiveness to skin for transdermal drug delivery. Advanced Functional Materials. 2020;**30**:1-10. DOI: 10.1002/adfm.202004407

[30] Admase AT, Fanta SW, Mersha DA. Development and characterization of sustainable biodegradable wood adhesive using starch as the main ingredient. Journal of the Indian Academy of Wood Science. 2024;**21**:332-344. DOI: 10.1007/s13196-024-00352-3

[31] Liu L, Jia Y, Zheng L, Luo R, Essawy H, Huang H, et al. Development and characterization of bio-based formaldehyde free sucrose-based adhesive for fabrication of plywood. Polymers (Basel). 2024;**16**:1-13. DOI: 10.3390/polym16050640

[32] Luo J, Zhou Y, Zhang Y, Gao Q. Polymers an eco-effective soybean meal-based adhesive. Polymers (Basel). 2020;**12**:640

[33] Ait Benhamou A, Boussetta A, Kassab Z, Nadifiyine M, Hamid Salim M, Grimi N, et al. Investigating the characteristics of cactus seeds by-product and their use as a new filler in phenol formaldehyde wood adhesive. International Journal of Adhesion and Adhesives. 2021;**110**:102940. DOI: 10.1016/j.ijadhadh.2021.102940

[34] Gadhave RV, Vineeth SK, Mahanwar PA, Pradeep T. Effect of addition of boric acid on thermo- mechanical properties of microcrystalline cellulose/polyvinyl alcohol blend and applicability as wood adhesive. Journal of Adhesion Science and Technology. 2020;**0**:1-15. DOI: 10.1080/01694243.2020.1832775

[35] Peng Z, Jiang X, Si C, Cárdenas-Oscanoa AJ, Huang C. Advances of modified lignin as substitute to develop lignin-based phenol-formaldehyde resin adhesives. ChemSusChem. 2023;**16**:e202300174

[36] Watcharakitti J, Win EE, Nimnuan J, Smith SM. Modified starch-based adhesives: A review. Polymers (Basel). 2022;**14**:2023. DOI: 10.3390/polym14102023

[37] Vineeth SK, Ravindra V, Gadhave PTG. Glyoxal cross-linked polyvinyl alcohol-microcrystalline cellulose blend as a wood adhesive with enhanced mechanical, thermal and performance properties. Materials International. 2020. DOI: 10.33263/Materials23.277285

[38] Perera KY, Hopkins M, Jaiswal AK, Jaiswal S. Nanoclays-containing bio-based packaging materials: Properties, applications, safety, and regulatory issues. Journal of Nanostructure in Chemistry. 2023;**14**:71-93. DOI: 10.1007/s40097-023-00525-5

[39] Beryl JR, Xavier JR. Halloysite for clay—Polymer nanocomposites: Effects of nanofillers on the anti-corrosion, mechanical, microstructure, and flame-retardant properties—A review. Journal of Materials Science. 2023;**58**:10943-10974. DOI: 10.1007/s10853-023-08710-1

[40] Necolau I, Damian CM, Claudiu R, Chiriac A, Mihail G, Vasile E, et al. Layered clay—Graphene oxide nanohybrids for the reinforcement and fire-retardant properties of polyurea matrix. Polymers (Basel). 2022;**14**:66

[41] Persano F, Leporatti S. Nano-clays for cancer therapy: State-of-the art and future perspectives. Journal of Personalized Medicine. 2022;**12**:1-26. DOI: 10.3390/jpm12101736

[42] Zhou Y, Zeng G, Zhang F, Tang Z, Luo J, Li K, et al. Preparation of functional fiber hybrid enhanced high strength and multifunctional protein based adhesive. Materials and Design. 2022;**224**:111289. DOI: 10.1016/j.matdes.2022.111289

[43] Yan N, Chen H, Gnanasekar P, Nair SS, Xu W, Chauhan P. Lignin as a key component in lignin-containing cellulose nanofibrils for enhancing the performance of polymeric diphenylmethane diisocyanate wood adhesives. ACS Sustainable Chemistry & Engineering. 2020;**8**:17165-17176. DOI: 10.1021/acssuschemeng.0c05642

[44] Abdoul-latif FM, El Montassir Z, Ainane A, Gharby S. Use of thymus plants as an ecological filler in urea-formaldehyde adhesives intended for bonding plywood. Processes. 2022;**10**:1-11

[45] Gavrilovi I, Teši T, Stupar S, Miloševi M. Bio-epoxy resins based on lignin and tannic acids as wood adhesives—Characterization and bonding properties. Polymers (Basel). 2024;**16**:2602

[46] Feng S, Shui T, Wang H, Ai X, Kuboki T, Xu CC. Properties of phenolic adhesives formulated with activated organosolv lignin derived from cornstalk. Industrial Crops and Products. 2021;**161**:113225. DOI: 10.1016/j.indcrop.2020.113225

[47] Mensah P, de Melo RR, Mitchual SJ, Owusu FW, Mensah MA, Donkoh MB, et al. Ecological adhesive based on cassava starch: A sustainable alternative to replace urea-formaldehyde (UF) in particleboard manufacture. Matéria (Rio Janeiro). 2024;**29**:e20230373. DOI: 10.1590/1517-7076-rmat-2023-0373

[48] Solihat NN, Santoso EB, Karimah A, Madyaratri EW, Sari FP, Falah F, et al. Physical and chemical properties of *Acacia mangium* lignin isolated from pulp mill byproduct for potential application in wood composites. Polymers (Basel). 2022;**14**:491. DOI: 10.3390/polym14030491

[49] Luo J, Luo J, Li X, Li K, Gao Q, Li J. Toughening improvement to a soybean meal-based bioadhesive using an interpenetrating acrylic emulsion network. Journal of Materials Science. 2016;**51**:9330-9341. DOI: 10.1007/s10853-016-0180-5

[50] Chen M, Zhang Y, Li Y, Shi SQ, Li J, Gao Q, et al. Soybean meal-based wood adhesive enhanced by phenol hydroxymethylated tannin oligomer for exterior use. Polymers (Basel). 2020;**12**:758

[51] Chen M, Luo J, Shi R, Zhang J, Gao Q, Li J. Improved adhesion performance of soy protein-based adhesives with a larch tannin-based resin. Polymers (Basel). 2017;**9**:408. DOI: 10.3390/polym9090408

[52] Yuan J, Du G, Yang H, Liu S, Wu Y, Ni K. Functionalization of cellulose with amine group and cross-linked with branched epoxy to construct high-performance wood adhesive. International Journal of Biological Macromolecules. 2022;**222**:2719-2728. DOI: 10.1016/j.ijbiomac.2022.10.053

[53] Zhu X, Song C, Sun X, Wang D, Cai D, Wang Z, et al. Improved water resistance of TA-modified soy adhesive: Effect of complexation. International Journal of Adhesion and Adhesives.

2021;**108**:102858. DOI: 10.1016/j.ijadhadh.2021.102858

[54] Zidanes UL, Dias MC, Lorenço MS, Araujo S, Júnnia M, Sousa TB, et al. Preparation and characterization of tannin-based adhesives reinforced with cellulose nano fibrils for wood bonding. Holzforschung. 2021;**75**:159-167

[55] Wang Z, Zhao S, Pang H, Zhang W, Zhang S, Li J. Developing eco-friendly high-strength soy adhesives with improved ductility through multiphase core-shell hyperbranched polysiloxane. ACS Sustainable Chemistry & Engineering. 2019;**7**:7784-7794. DOI: 10.1021/acssuschemeng.8b06810

[56] Pradyawong S, Qi G, Li N, Sun XS, Wang D. Adhesion properties of soy protein adhesives enhanced by biomass lignin. International Journal of Adhesion and Adhesives. 2017;**75**:66-73. DOI: 10.1016/j.ijadhadh.2017.02.017

[57] Liu L-Y, Wan-Shuan Chiang HC, Yeh T-F. Phenolation to improve hardwood kraft lignin for wood. Polymers (Basel). 2024;**16**:1923

[58] Jin S, Li K, Gao Q, Zhang W, Chen H, Li J, et al. Multiple crosslinking strategy to achieve high bonding strength and antibacterial properties of double-network soy adhesive. Journal of Cleaner Production. 2020;**254**:120143. DOI: 10.1016/j.jclepro.2020.120143

[59] Sain S, Matsakas L, Rova U, Christakopoulos P, Öman T, Skrifvars M. Spruce Bark-extracted lignin and tannin-based bioresin-adhesives: Effect of curing temperatures on the thermal properties of the resins. Molecules. 2021;**26**:3523

[60] Chang Z, Pang H, Huang A, Li J, Zhang S. Reinforcement of bonding strength and water resistance of soybean meal-based adhesive via construction of an interactive network from biomass residues. Polymers (Basel). 2019;**11**:967. DOI: 10.3390/polym11060967

[61] Chang M, Liu X, Wang X, Peng F, Ren J. Mussel-inspired adhesive hydrogels based on biomass-derived xylan and tannic acid cross-linked with acrylic acid with antioxidant and antibacterial properties. Journal of Materials Science. 2021;**56**:14729-14740. DOI: 10.1007/s10853-021-06228-y

[62] Samyn P, Adeel M, Pazdur L, Billen P. (Thermo)mechanical and chemical characteristics of photochemically crosslinked acrylates from bio- and fossil-based origin. Journal of Polymer Science. 2024;**62**:3511-3527. DOI: 10.1002/pol.20230969

[63] Osman Z. Thermomechanical analysis of the tannins of Acacia Nilotica spp. Nilotica as a rapid tool for the evaluation of wood-based adhesives. Journal of Thermal Analysis and Calorimetry. 2012;**107**:709-716. DOI: 10.1007/s10973-011-1721-4

[64] Arias A, González-Rodríguez S, Vetroni Barros M, Salvador R, de Francisco AC, Moro Piekarski C, et al. Recent developments in bio-based adhesives from renewable natural resources. Journal of Cleaner Production. 2021;**314**:127892. DOI: 10.1016/j.jclepro.2021.127892

[65] Wu J, Yu X, Zhang H, Guo J, Hu J, Li M-H. Fully biobased vitrimers from glycyrrhizic acid and soybean oil for self-healing, shape memory, weldable, and recyclable materials. ACS Sustainable Chemistry & Engineering. 2020;**8**:6479-6487. DOI: 10.1021/acssuschemeng.0c01047

[66] Gadhave RV, Srivastava S, Mahanwar PA, Gadekar PT. Lignin:

Renewable raw material for adhesive. Open Journal of Polymer Chemistry. 2019;**09**:27-38. DOI: 10.4236/ojpchem.2019.92003

[67] Moubarik A, Pizzi A, Allal A, Charrier F, Charrier B. Cornstarch and tannin in phenol-formaldehyde resins for plywood production. Industrial Crops and Products. 2009;**30**:188-193. DOI: 10.1016/j.indcrop.2009.03.005

[68] Dhawale PV, Vineeth SK, Gadhave RV, Jabeen Fatima MJ, Supekar MV, Thakur VK, et al. Tannin as a renewable raw material for adhesive applications: A review. Materials Advances. 2022;**3**:3365-3388. DOI: 10.1039/d1ma00841b

[69] Gouveia JR, Antonino LD, Garcia GES, Tavares LB, Santos ANB, dos Santos DJ. Kraft lignin-containing polyurethane adhesives: The role of hydroxypropylation on thermomechanical properties. The Journal of Adhesion. 2021;**97**:1423-1439. DOI: 10.1080/00218464.2020.1784148

[70] Martinez-Sola C, Garcia-Teruel P, Martinez-Solano P. Characterization of a formaldehyde-free cornstarch-tannin wood adhesive for interior plywood. European Journal of Wood and Wood Products. 2010;**46**:427-433

[71] Chen X, Xi X, Pizzi A, Fredon E, Du G, Gerardin C, et al. Oxidized demethylated lignin as a bio-based adhesive for wood bonding. The Journal of Adhesion. 2020;**00**:1-18. DOI: 10.1080/00218464.2019.1710830

[72] Li R, Li W, Zheng F, Zhang Y, Hu J. Versatile bio-based epoxy resin: From banana waste to applied materials. Journal of Applied Polymer Science. 2019;**136**:1-8. DOI: 10.1002/app.47135

[73] Magalhães S, Alves L, Medronho B, Fonseca AC, Romano A, Coelho JFJ, et al. Brief overview on bio-based adhesives and sealants. Polymers (Basel). 2019;**11**:1685. DOI: 10.3390/polym11101685

[74] Moubarik A, Allal A, Pizzi A, Charrier F, Charrier B. Preparation and mechanical characterization of particleboard made from maritime pine and glued with bio-adhesives based on cornstarch and tannins. Maderas. Ciencia y Tecnología. 2010;**12**:189-197. DOI: 10.4067/S0718-221X2010000300004

[75] Jalali S, da Silva Pereira Borges C, RJC C, de Sousa Marques EA, Bordado JCM, da Silva LFM. Characterization of densified pine wood and a zero-thickness bio-based adhesive for eco-friendly structural applications. Materials (Basel). 2023;**16**:7147. DOI: 10.3390/ma16227147

[76] Muñoz-Gimena PF, Oliver-Cuenca V, Peponi L, López D. A review on reinforcements and additives in starch-based composites for food packaging. Polymers (Basel). 2023;**15**:2972. DOI: 10.3390/polym15132972

[77] Tenorio-Alfonso A, Sánchez MC, Franco JM. Synthesis and mechanical properties of bio-sourced polyurethane adhesives obtained from castor oil and MDI-modified cellulose acetate: Influence of cellulose acetate modification. International Journal of Adhesion and Adhesives. 2019;**95**:102404. DOI: 10.1016/j.ijadhadh.2019.102404

[78] Din Z, Chen L, Xiong H, Wang Z, Ullah I, Lei W. Starch: An undisputed potential candidate and sustainable resource for the starch: An undisputed potential candidate and sustainable resource for the development of wood adhesive. Starch-Stärke. 2020;**72**:1900276. DOI: 10.1002/star.201900276

[79] Zheng L, McClelland DJ, Rehmann KMS, Barnett KJ, Huber GW, Klier J. Bio-based 1,5-pentanediol as a replacement for petroleum-derived 1,6-hexanediol for polyester polyols, coatings, and adhesives. ACS Sustainable Chemistry & Engineering. 2022;**10**:5781-5791. DOI: 10.1021/acssuschemeng.1c08198

[80] Gumowska A, Kowaluk G. Physical and mechanical properties of high-density fiberboard bonded with bio-based adhesives. Forests. 2023;**14**:84

[81] Footprints E. Environmental impacts of traditional and innovative forest-based bioproducts. 2016

[82] Jarensungnen C, Jetsrisuparb K, Phanthanawiboon S, Theerakulpisut S, Hiziroglu S, Knijnenburg JTN, et al. Development of eco-friendly antifungal and antibacterial adhesive derived from modified cassava starch waste/polyvinyl alcohol containing green synthesized nano-silver. Scientific Reports. 2023;**13**:1-13. DOI: 10.1038/s41598-023-40305-3

[83] Tenorio-Alfonso A, Sánchez MC, Franco JM. Preparation, characterization and mechanical properties of bio-based polyurethane adhesives from isocyanate-functionalized cellulose acetate and castor oil for bonding wood. Polymers (Basel). 2017;**9**:1-14. DOI: 10.3390/polym9040132

[84] Segovia F, Blanchet P, Auclair N, Essoua GGE. Thermo-mechanical properties of a wood fiber insulation board using a bio-based adhesive as a binder. Buildings. 2020;**10**:1-15. DOI: 10.3390/buildings10090152

[85] Mo X, Zhang X, Fang L. Research Progress of Wood-Based Panels Made of Thermoplastics as Wood Adhesives. Basel, Switzerland: MDPI; 2022. pp. 1-12

[86] Wang XL, Chen L, Wu JN, Fu T, Wang YZ. Flame-retardant pressure-sensitive adhesives derived from epoxidized soybean oil and phosphorus-containing dicarboxylic acids. ACS Sustainable Chemistry & Engineering. 2017;**5**:3353-3361. DOI: 10.1021/acssuschemeng.6b03201

[87] Costa CSMF, Fonseca AC, Moniz J, Godinho M, Serra AC, Coelho JFJ. Soybean and coconut oil based unsaturated polyester resins: Thermomechanical characterization. Industrial Crops and Products. 2016;**85**:403-411. DOI: 10.1016/j.indcrop.2016.01.030

[88] Deka H, Karak N. Bio-based hyperbranched polyurethane/clay nanocomposites: Adhesive, mechanical, and thermal properties. Polymers for Advanced Technologies. 2011;**22**:973-980. DOI: 10.1002/pat.1603

[89] Wang L, Zhang X, Yang K, Fu YV, Xu T, Li S, et al. A novel double-crosslinking-double-network design for injectable hydrogels with enhanced tissue adhesion and antibacterial capability for wound treatment. Advanced Functional Materials. 2020;**30**:1-14. DOI: 10.1002/adfm.201904156

[90] Boussetta A, Ait A, Charii H, Houssaine E, Mehdi A. Formulation and characterization of chitin-starch bio-based wood adhesive for the manufacturing of formaldehyde-free composite particleboards. Waste and Biomass Valorization. 2023;**14**:3671-3687. DOI: 10.1007/s12649-023-02091-x

[91] Yang J, Wang S. Polysaccharide-based multifunctional hydrogel bio-adhesives for wound healing: A review. Gels. 2023;**9**:138

[92] Bashir Z, Yu W, Xu Z, Li Y, Lai J, Li Y, et al. Engineering bio-adhesives

based on protein—Polysaccharide phase separation. International Journal of Molecular Sciences. 2022;**23**:9987

[93] Tserpes K, Tzatzadakis V. Synthesis and experimental characterization of a MWCNT-filled bio-based adhesive. Aerospace. 2021;**8**:1-16. DOI: 10.3390/aerospace8020026

[94] Adelka YF, Purnomo D, WidyaningrumBA,SutiawanJ,UmemuraK, Hermawan D, et al. Effectiveness of nanocatalyst in the improvement of sorghum bagasse particleboard bonded with bio-adhesive. Jurnal Sylva Lestari. 2023;**11**:382-395

[95] Stoeckel F, Konnerth J, Gindl-altmutter W. Mechanical properties of adhesives for bonding wood—A review. International Journal of Adhesion and Adhesives. 2013;**45**:32-41. DOI: 10.1016/j.ijadhadh.2013.03.013

[96] Feng G, Djordjevic I, Mogal V, O'Rorke R, Pokholenko O, Steele TWJ. Elastic light tunable tissue adhesive dendrimers. Macromolecular Bioscience. 2016;**16**:1072-1082. DOI: 10.1002/mabi.201600033

[97] Andreas Öchsner RDA. Handbook of Adhesion Technology. Germen: Springer Science & Business Media; 2011

[98] Zeng Z, Mo X. Rapid in situ cross-linking of hydrogel adhesives based on thiol-grafted bio-inspired catechol-conjugated chitosan. Journal of Biomaterials Applications. 2017;**32**:612-621. DOI: 10.1177/0885328217738403

[99] Yang Z, Huang R, Zheng B, Guo W, Li C, He W, et al. Highly stretchable, adhesive, biocompatible, and antibacterial hydrogel dressings for wound healing. Advancement of Science. 2021;**8**:1-12. DOI: 10.1002/advs.202003627

[100] Borges CSP, Jalali S, Tsokanas P, Marques EAS, Carbas RJC, da Silva LFM. Sustainable development approaches through wooden adhesive joints design. Polymers (Basel). 2023;**15**:89. DOI: 10.3390/polym15010089

[101] Heinrich LA, Heinrich LA. Future opportunities for bio-based adhesives—Advantages beyond renewability. Green Chemistry. 2019;**21**:1866-1888. DOI: 10.1039/c8gc03746a

[102] Borges CSP, Akhavan-Safar A, Tsokanas P, Carbas RJC, Marques EAS, da Silva LFM. From Fundamental Concepts to Recent Developments in the Adhesive Bonding Technology: A General View. Berlin, Germen: Springer International Publishing; 2023

[103] Tian W, Wang X, Ye Y, Wu W, Wang Y, Jiang S, et al. Recent progress of biomass in conventional wood adhesives: A review. Green Chemistry. 2023;**25**:10304-10337. DOI: 10.1039/d3gc03441k

[104] Gadhave RV, Mahanwar PA, Gadekar PT. Starch-based adhesives for wood/wood composite bonding: Review. Open Journal of Polymer Chemistry. 2017;**7**:19-32. DOI: 10.4236/ojpchem.2017.72002

[105] Rowell RM. Chemical modification of wood for improved adhesion in composites. Proceedings of Wood Adhesives. 1995;**7296**:56-60

[106] Jouki M, Tabatabaei Yazdi F, Mortazavi SA, Koocheki A. Physical, barrier and antioxidant properties of a novel plasticized edible film from quince seed mucilage. International Journal of Biological Macromolecules. 2013;**62**:500-507. DOI: 10.1016/j.ijbiomac.2013.09.031

[107] Sridach W, Hodgson KT, Nazhad MM. Biodegradation and

recycling potential of barrier coated paperboards. BioResources. 2007;**2**:179-192. DOI: 10.15376/biores.2.2.179-192

[108] Eissenberger K, Ballesteros A, De Bisschop R, Bugnicourt E, Cinelli P, Defoin M, et al. Approaches in sustainable, biobased multilayer packaging solutions. Polymers. 2023;**15**:1184

[109] Liu Z, Song Z, Lv B, Qiu Z. Rapid Reassembly, Biomass-Derived Adhesive Based on. Basel, Switzerland: MDPI; 2023. pp. 1-11

[110] Antov P, Savov V, Neykov N. Sustainable bio-based adhesives for eco-friendly wood composites a review. Wood Research. 2020;**65**:51-62. DOI: 10.37763/wr.1336-4561/65.1.051062

[111] Jawaid M, Khan TA, Nasir M, Asim M. Eco-Friendly Adhesives for Wood and Natural Fiber Composites Characterization, Fabrication and Applications. Composites Science and Technology. Singapore: Springer; 2021

[112] Yu C, Chen Y, Li R, Jiang J. A Narrative Review: Modification of Bio-Based Wood Adhesive for Performance Improvement. Basel, Switzerland: MDPI; 2024. pp. 1-13

[113] Guadagno L, Sarno M, Vietri U, Raimondo M, Cirillo C, Ciambelli P. Graphene-based structural adhesive to enhance adhesion performance. RSC Advances. 2015;**5**:27874-27886. DOI: 10.1039/c5ra00819k

[114] Maulana S, Wibowo ES, Mardawati E, Iswanto AH, Papadopoulos A, Adly M, et al. Eco-Friendly and High-Performance Bio-Polyurethane Adhesives from Vegetable Oils: A Review. Basel, Switzerland: MDPI; 2024. pp. 1-29

[115] Ayoola F. Production of adhesive from cassava starch. Journal of Physics: Conference Series. 2019;**1378**:032079. DOI: 10.1088/1742-6596/1378/3/032079

Chapter 3

Next-Generation Bonding Agents for Sustainable FRP Retrofitting of Concrete Structures

Mohammad Al-Zu'bi

Abstract

This chapter explores the advancements in nanomaterial-modified epoxy adhesives (NMEAs) for structural retrofitting, focusing on the integration of various carbon-based nanomaterials, including carbon nanofibers (CNFs), cellulose nanocrystals (CNCs), and graphite nanopowder, as well as silicon-based nanomaterials such as silica nanopowder and montmorillonite (MMT) nanoclay. These nanomaterials are incorporated into epoxy matrices to enhance mechanical, thermal, and bonding properties. Advanced characterization techniques, including scanning electron microscopy (SEM), X-ray diffraction (XRD), Fourier transform infrared (FTIR), and Raman spectroscopies, are employed to analyze nanomaterial dispersion through the polymer matrix, crystallinity, and chemical interactions. Experimental results demonstrate the environmental benefits and increased durability of NMEAs, establishing them as sustainable and efficient alternatives to neat epoxy (NE) adhesives in fiber-reinforced polymer (FRP) retrofitting systems. By improving mechanical strength, bonding efficiency, and long-term resilience, NMEAs offer a transformative approach to structural retrofitting. This chapter integrates experimental findings and practical applications to enhance the understanding of adhesive modifications and their critical role in promoting sustainable construction practices. The chapter concludes with recommendations for future research directions aimed at filling the research and knowledge gaps in the field of advancing the development and application of FRP-retrofitting techniques using the NMEAs.

Keywords: nanomaterial-modified epoxy adhesives (NMEAs), fiber-reinforced polymer (FRP) composites, structural retrofitting, adhesive modifications, sustainable construction materials, polymer nanocomposites, carbon-based nanomaterials, silicon-based nanomaterials

1. Introduction

1.1 Overview of structural retrofitting and the role of adhesives

The success of retrofitting concrete structures using fiber-reinforced polymer (FRP) materials is strongly influenced by the choice of bonding agent, which is responsible for securing the FRP reinforcement to the concrete substrate.

This adhesive layer acts as a stress-transfer medium between the FRP and the concrete. Consequently, the properties of the adhesive are critical in improving the bond strength and performance at the FRP-adhesive-concrete interfaces, making the selection of appropriate bonding agents essential for achieving effective retrofitting outcomes [1–3].

Epoxy is widely regarded as one of the most significant thermosetting polymers for retrofitting applications due to its exceptional thermal and mechanical stability, outstanding chemical resistance, electrical insulating properties, and lightweight nature. As a result, neat epoxy (NE) adhesives have proven effective in concrete retrofitting for both externally bonded reinforcement (EBR) and near surface-mounted (NSM) systems. These adhesives facilitate the bonding of FRP composites to existing concrete substrates, thereby enhancing the strength, stiffness, and overall structural integrity of the retrofitted members [2–7].

Despite their advantages, the use of FRP composites for retrofitting concrete elements presents certain challenges related to the performance of the epoxy adhesive layer. Issues such as the debonding of FRP materials from the concrete substrate can occur, often due to non-uniform stress distribution at the concrete-adhesive and FRP-adhesive interfaces. This non-uniformity can lead to premature debonding failures in FRP-retrofitted concrete structures [8–11]. Additionally, highly cross-linked epoxy resins are inherently rigid and brittle, with limited resistance to crack initiation and propagation. These shortcomings not only restrict their applicability in structural retrofitting [5, 11] but also adversely impact the durability and safety of retrofitted structures. To address these limitations, the performance of epoxy adhesives must be improved for effective application in both EBR-FRP and NSM-FRP systems. Incorporating nanoparticles into epoxy formulations has emerged as a promising solution, significantly enhancing the mechanical, thermal, and bonding properties of epoxy adhesives.

1.2 Scope and objectives of the chapter

This chapter provides a comprehensive overview of the advancements in nanomaterial-modified epoxy adhesives (NMEAs) for structural retrofitting applications, emphasizing their potential to revolutionize the performance and sustainability of FRP systems used in concrete retrofitting. The scope encompasses the integration of carbon- and silicon-based nanomaterials into epoxy adhesives, exploring their role in enhancing critical properties such as mechanical strength, thermal stability, and bonding efficiency.

A key objective of this chapter is to examine the characteristics influenced by the incorporation of nanomaterials, including improvements in the mechanical, thermal, chemical, and microstructural properties of NE adhesives.

The chapter also discusses how these advancements contribute to increased durability and environmental sustainability, positioning NMEAs as efficient alternatives to traditional adhesives in retrofitting applications.

Additionally, the chapter integrates experimental findings and practical applications to provide insights into the performance of NMEAs under diverse environmental and operational conditions. By identifying areas for future research and development, this work also aims to promote the adoption of NMEAs in sustainable construction practices, advancing the field of structural retrofitting and ensuring long-term infrastructure resilience.

2. Effect of incorporating nanomaterials on the properties of epoxy adhesives

2.1 Mechanical and thermal properties

It was found that incorporating carbon-based nanomaterials, such as carbon nanotubes (CNTs) [1, 2, 10, 12, 13], carbon nanofibers (CNF) [14–18], graphene and graphite [19–25], and cellulose nanocrystals (CNCs) [26–31] into epoxy adhesives, improved their mechanical and thermal properties and were also found to provide outstanding reinforcing potential and demonstrate efficient stress transfer behavior. The enhancements in properties are primarily due to the strong interfacial interactions established between the nanoparticles and the epoxy matrix through chemical reactions. Moreover, achieving uniform nanoparticle dispersion within the epoxy facilitates a robust bond between the nanoparticles and unreacted functional groups of the epoxy, effectively arresting and suppressing crack propagation. This, in turn, significantly improves the adhesive's overall performance. For example, Ashrafi et al. [10] demonstrated that adding 0.1 wt.% of single-walled carbon nanotubes (SWCNTs) to epoxy resulted in a 13% and 28% increase in mode I and mode II interlaminar fracture toughness, respectively, alongside a 3.5% improvement in compression-after-impact (CAI) strength compared to the NE. Additionally, the incorporation of SWCNTs reduced the area of impact damage by 5%. The observed increase in fracture toughness was attributed to crack bridging mechanisms, including pull-out and peeling, in Mode I, as well as the formation of larger hackles in Mode II. Similarly, Iwahori et al. [14] reported that the addition of 10 wt.% cup-stacked carbon nanofibers (CSNF) enhanced tensile strength and modulus by 17% and 45%, respectively, compared to the NE. Furthermore, flexural strength and modulus exhibited improvements of 38% and 42%, respectively. These enhancements were linked to the unique microstructure of CSNF's cup-stacked geometry, which likely inhibits graphene sheet sliding.

Silicon-based nanomaterials, such as silica nanopowder [8, 32–34] and montmorillonite (MMT) nanoclay [35, 36], have also showed their efficiency in improving the mechanical and thermal performance of epoxy polymers. For instance, Chisholm et al. [33] reported that incorporating 1.5 wt.% of SiC nanoparticles into an epoxy matrix significantly enhanced its mechanical properties, including tensile and flexural strengths, tensile and flexural moduli, and stiffness, with improvements of 11.6%, 39.2%, 23.5%, 12.4%, and 12%, respectively. These enhancements were attributed to the uniform distribution of nanoparticles throughout the matrix, which reduced void content in the composites by approximately 30–40%, as observed in tests conducted on the nano-phased composites. The SiC nanoparticles effectively acted as fillers, occupying void spaces and contributing to the overall structural integrity of the matrix. Similarly, Hosur et al. [35] observed that introducing 2 wt.% of MMT into an epoxy resin increased the storage modulus and loss modulus by 12% and 6%, respectively, while raising the glass transition temperature (T_g) by 14.54°C. These improvements were linked to the homogeneous dispersion of MMT particles within the epoxy matrix, which enhanced its thermal and mechanical performance.

2.2 Chemical properties

The incorporation of nanoparticles into polymer matrices has been shown to significantly affect their chemical structure. As a result, it is crucial to identify the chemical reactions and compositional changes occurring within the matrix due to the inclusion of nanoparticles. Techniques such as Fourier transform infrared (FTIR) and Raman spectroscopies are widely utilized to achieve this. These methods provide valuable insights into the molecular structure of materials by detecting functional groups that emerge or disappear following the addition of nanomaterials. This analysis enables researchers to identify the characteristic changes within the polymer, offering a deeper understanding of its modified properties.

2.2.1 FTIR spectroscopy

Few research studies [36–40] have explored the impact of nanoparticles on the chemical composition of epoxy, specifically changes in functional groups. For example, Huttunen-Saarivirta et al. [36] reported that while clay nanoparticles in the epoxy matrix did not introduce new peaks, the positions of certain characteristic IR peaks shifted. Further research [37] revealed that dispersing 0.5 wt.% ultra-sonicated ozonolytic (USO)-treated CNTs (OZ-CNTs) in DGEBA epoxy resin led to the formation of an esters peak at 1754 cm^{-1}, indicating a reaction between OZ-CNTs and the epoxy matrix. Wang et al. [38] analyzed FTIR data for epoxy specimens modified with CNT, GNP, and fullerene-C60 and found identical characteristic peaks across all samples, suggesting no new chemical bonds formed. However, the disappearance of the peak at 913 cm^{-1} indicated ring-opening polymerization during the curing process. Montazeri [39] observed no additional bands when 2 wt.% MWNT-OH was mixed with DGEBA epoxy. Similarly, Al-Zu'bi et al. [40] found that carbon-based materials, including CNF, graphite, and CNCs, generally did not form new chemical bonds in epoxy, except for 1 wt.% CNF, which showed a specific bond (C═O stretching at 1710 cm^{-1}). Conversely, silicon-based materials, such as silica nanopowder and MMT nanoclay, did not alter the chemical bonds in NE, with no new bonds forming or existing ones disappearing.

2.2.2 Raman spectroscopy

Raman spectroscopy has been widely employed in numerous studies [23, 37, 40–51] to examine the chemical structure of epoxy polymers and their nanocomposites. A key aspect of Raman analysis for carbon-based materials is the intensity ratio of the D and G bands (I_D/I_G), which serves as an indicator of structural defects in carbon nanomaterials. The G band arises from in-plane vibrations of sp^2-bonded carbon atoms, while the D band is associated with out-of-plane vibrations due to structural imperfections. For instance, research [41] identified D and G bands at 1358 cm^{-1} and 1593 cm^{-1}, respectively, in polyphosphamide (PPA)/epoxy and PPA-GNS/epoxy specimens. Zhao et al. [43] found minimal changes in the D/G intensity ratio for single-walled CNTs (SWCNTs) before ($I_D/I_G = 0.09$) and after epoxy adsorption (EpPy-16) ($I_D/I_G = 0.11$). Similarly, the Raman spectra of GNS-based epoxy resins [47] showed D-band peaks at 1250–1450 cm^{-1} and G-band peaks at 1500–1700 cm^{-1}, reflecting the dispersion of graphene nanosheets (GNS) in the epoxy matrix. In the work of Al-Zu'bi et al. [40], Raman results demonstrated higher I_D/I_G ratios for CNF and graphite NMEAs compared to their pristine counterparts. For CNF NMEAs, D-band intensity followed the order 0.5 > 1.0 > 1.5 (wt.%), whereas G-band intensity exhibited the reverse order of 0.5 < 1.0 < 1.5. A similar trend was observed in graphite NMEAs, where D-band intensity followed the sequence 0.5 < 1.5 < 1.0.

2.3 Microstructural properties

2.3.1 Crystallinity

First, it is worth noting that crystallinity is generally considered a microstructural property of materials because it describes the arrangement of atoms or molecules within a material on a microscopic scale. Specifically, it refers to the degree of structural order in a material, such as the proportion of a solid that is crystalline versus amorphous.

Determining the degree of crystallinity in polymer nanocomposites is essential for predicting their properties and identifying potential applications. X-ray diffraction (XRD) is widely employed for crystallinity analysis. Studies have shown that nanoparticles within a polymer matrix can either act as nucleating agents, enhancing the composite's crystallinity, or as impurities, disrupting the formation of ordered structures [40, 52–62]. Mahmood et al. [53] observed that introducing amine-functionalized CNTs into polyamide 6 (PA6) increased the crystallinity of the nanocomposites. Similarly, Morimune et al. [54] reported that incorporating graphene oxide (GO) into polyvinyl alcohol (PVA) elevated its crystallinity from 28% in pure PVA to 30%, 33%, 35%, and 39% for GO loadings of 0.1%, 0.5%, 1%, and 5%, respectively. Conversely, Bhattacharyya et al. [55] found a reduction in crystallinity in graphene-reinforced ultrahigh molecular weight polyethylene (UHMWPE). Werner et al. [56] also noted that adding silica nanoparticles to PA6 decreased its crystallinity, with further reductions observed as silica concentration increased. Similar trends were identified in ethylene vinyl acetate (EVA)-silica nanocomposites [57] and silica-PVA nanocomposites [58]. However, Peng et al. [59] found that at low silica loading (0.5 wt.%), PVA's crystallinity improved, attributed to silica acting as a heterogeneous nucleating agent. Kumar et al. [60] investigated PVA-cellulose nanocomposites and observed a reduction in crystallinity. This reduction was explained by the strong interactions between CNCs and PVA's hydroxyl (—OH) groups, which formed a tangled mass, creating steric hindrance and disrupting the highly ordered arrangement of PVA.

2.3.2 Other microstructural properties

Scanning electron microscopy (SEM) is widely used to analyze the microstructure of epoxy nanocomposites and assess the dispersion of nanoparticles within the polymer matrix [13, 23, 28, 36–43, 45, 47–49, 63–67]. Although limited, some studies have investigated the porosity of polymers [40, 68, 69]. For instance, research [40] indicated that nanoparticle agglomerations, including densely packed regions, contributed to increased porosity in nanocomposites. The porosity levels were observed to rise with increasing nanoparticle concentrations, with a more significant increase when transitioning from 0.5 to 1.0 wt.% compared to the transition from 1.0 to 1.5 wt.%. Furthermore, the porosity trends within carbon-based nanocomposites varied with concentration: From 0.5 to 1.0 wt.%, the order was "CNF < cellulose < graphite," but from 1.0 to 1.5 wt.%, it shifted to "CNF < graphite < cellulose." On the other hand, silicon-based nanocomposites exhibited consistent porosity behavior across all concentration ranges, with clay consistently resulting in a higher % porosity increase than silica (clay ≫ silica). These findings highlight the intricate and concentration-dependent behavior of carbon-based systems, whereas silicon-based nanocomposites demonstrated more predictable and consistent trends/behavior in porosity reduction.

3. Applications of NMEAs for FRP-structural retrofitting

As mentioned previously, thanks to their exceptional thermo-mechanical properties, the incorporation of nanoparticles into epoxy adhesives has proven effective in addressing the limitations associated with the use of NE adhesives in both EBR- and NSM-FRP retrofitting of concrete members through enhancing their mechanical properties and interfacial bond between FRP reinforcement and concrete substrate [1, 2, 12, 13, 70–73].

3.1 Mechanical performance

3.1.1 EBR-FRP retrofitting technique

Irshidat et al. [1] investigated the impact of using both neat epoxy (NE) and CNTs-modified epoxy on the flexural performance of reinforced concrete (RC) beams strengthened with carbon FRP (CFRP) sheets. The study revealed that incorporating CNTs into epoxy resulted in a modest 5% increase in the ultimate load capacity of the beams, while significantly enhancing their stiffness and toughness by approximately 35% and 28%, respectively. The modified epoxy was also observed to delay the propagation and debonding of CFRP sheets, attributed to improved concrete-epoxy adhesion. Similarly, a separate study [12] examined the application of CNTs-modified epoxy in strengthening RC columns and reported a 12% increase in axial load capacity and a 19% improvement in toughness compared to columns bonded with NE. Both NE- and CNTs-modified epoxy specimens exhibited ductile failure modes. Rousakis et al. [13] explored the use of multi-walled CNTs (MWCNTs) in repairing and strengthening epoxy resin systems for FRP-confined concrete cylinders. Their findings underscored the effectiveness of MWCNTs-reinforced epoxy resins in advanced crack repair for high-performance concrete, demonstrating the material's potential for enhancing structural integrity.

3.1.2 NSM-FRP retrofitting technique

To the best of the authors' knowledge, the use of NMEAs in NSM-FRP retrofitting applications has been explored in only one study to date [74]. In this investigation, concrete prisms were subjected to three-point bending tests to evaluate their load-carrying capacities, ductility responses, and structural failure modes. The findings revealed that compared to specimens-bonded NE, the use of NMEAs incorporating CNF and cellulose led to reductions in flexural capacities by approximately 37% and 9%, respectively. These reductions were attributed to weak interfacial bonding caused by particle agglomeration and the high porosity of the resulting nanocomposites. Despite this, significant improvements in ductility were observed, with increases of about 49% and 36%, respectively, demonstrating the potential of NMEAs to enhance the deformation capacity of retrofitted concrete elements.

The study further revealed that specimens bonded with silica-, clay-, and graphite-modified epoxies exhibited capacity enhancements of approximately 17%, 5%, and 15%, respectively, compared to those bonded with NE. However, the increased crystallinity of graphite nanocomposites resulted in a 12% reduction in ductility for the graphite-epoxy-bonded specimens. In contrast, silica and clay nanocomposites demonstrated ductility enhancements of about 19% and 24%, respectively.

Additionally, the use of carbon-based NMEAs prevented interfacial debonding and altered the failure mode, as observed in the specimens bonded with 0.1 wt.%

CNF-modified epoxy from shear to flexural failure. On the other hand, slight debonding at the bar-adhesive interface was noted in specimens retrofitted with silicon-based NMEAs, such as silica nanocomposites, which was accompanied by cohesive failure in the adhesive layer and minor concrete crushing. Nevertheless, the concrete-adhesive interfacial bonding remained intact, and the shear failure mode was preserved.

3.2 Bond behavior

Irshidat and Al-Saleh [2] examined the effect of CNTs-modified epoxy on the bond-slip behavior between the concrete surface and carbon or glass FRP sheets in concrete prisms. Their findings indicated that incorporating CNTs into epoxy resin improved the bond strength and slip at failure by 35% and 52% for carbon fiber, and by 26% and 83% for glass fiber, respectively. Additionally, while NE-bonded specimens exhibited debonding failure at the interface between the fiber sheet and the concrete surface, those bonded with CNTs-modified epoxy experienced cohesive failure within the adhesive layer.

The bond characteristics and strength of cement paste (CP)-adhesive joints bonded with NMEAs were evaluated using a lap-shear test [73]. The findings revealed that incorporating 0.5 and 1.0 wt.% of carbon-based nanofillers, except for 1.0 wt.% CNF, significantly improved the joint strength. However, further loading at 1.5 wt.% led to a reduction in strength, with values dropping below those of NE-bonded joints. Conversely, all joints bonded with epoxy containing silicon-based nanofillers exhibited increased strength compared to unfilled epoxy, regardless of the loading percentage. The strength of the joints decreased as the nanofiller concentration increased, attributed to particle agglomeration. A mixed cohesive and adhesive failure mode was observed in all tested joints.

4. Conclusions

This chapter highlights the transformative potential of nanomaterial-modified epoxy adhesives (NMEAs) in advancing structural retrofitting practices. By incorporating carbon- and silicon-based nanomaterials, these adhesives significantly enhance mechanical, thermal, and bonding properties, addressing the limitations of traditional adhesives. The improved performance of NMEAs stems from their ability to achieve better dispersion, increased crystallinity, and stronger chemical interactions within the adhesive matrix, as demonstrated through various experimental investigations.

The use of NMEAs in fiber-reinforced polymer (FRP) systems for concrete retrofitting has shown remarkable improvements in mechanical behavior, bonding efficiency, and durability performance. These properties contribute to long-term structural resilience and make NMEAs sustainable and efficient alternatives to conventional epoxy adhesives. Furthermore, the environmental benefits of NMEAs, including reduced material waste and extended infrastructure lifespan, align with global efforts toward sustainable construction practices.

5. Recommendations for future work

Based on the research studies discussed in this chapter regarding the utilization of the NMEAs in retrofitting applications for concrete members, it is evident that while significant progress has been made, several gaps remain unaddressed. These gaps provide opportunities for researchers to further explore and expand this domain, ultimately enhancing best practices and optimizing repair performance.

It has been demonstrated that the use of NMEAs, rather than neat adhesives, effectively enhances the mechanical and bond performance of retrofitted concrete elements. However, investigations in this field remain limited. To bridge this gap, further experimental and numerical studies are highly recommended to evaluate the efficiency of NMEAs under various loading and environmental conditions. These efforts should focus on their long-term performance and resilience, especially in applications involving FRP retrofitting systems.

For further insights into the potential of nanomaterials to improve the mechanical and thermal properties of both epoxy and cement-based adhesives, researchers are advised to explore the detailed review provided in Ref. [3]. Additionally, significant findings are presented in the pioneering study [74] (discussed in Section 3.1.2), which examines the application of NMEAs in NSM-FRP retrofitting of concrete elements, highlighting their promising role in enhancing concrete rehabilitation. These works offer valuable resources for advancing the use of nanotechnology in structural repair. Researchers may also refer to comprehensive reviews [75, 76] on EBR- and NSM-FRP retrofitting techniques, which provide essential insights and methodologies relevant to CFRP repair of heat-damaged RC beams.

Acknowledgements

The author would like to acknowledge that some sections of this chapter are derived from the doctoral thesis titled *"Nanotechnology-Enhanced Near-Surface Mounted-Fibre Reinforced Polymers (NSM-FRP) Structural Retrofitting"*, authored by Mohammad Al-Zu'bi, Mizi Fan, and Lorna Anguilano, and archived in the Brunel University Research Archive (BURA). The thesis is available at the following link: https://bura.brunel.ac.uk/handle/2438/29037.

The author also expresses gratitude to Brunel University of London for the support provided during the doctoral research that contributed to the foundation of this chapter.

Conflict of interest

The author declares no conflict of interest.

Author details

Mohammad Al-Zu'bi
Department of Civil and Environmental Engineering, College of Engineering, Design and Physical Sciences, Brunel University of London, United Kingdom

*Address all correspondence to: 1939105@alumni.brunel.ac.uk

References

[1] Irshidat MR, Al-Saleh MH, Almashagbeh H. Effect of carbon nanotubes on strengthening of RC beams retrofitted with carbon fiber/epoxy composites. Materials and Design. 2016;**89**:225-234

[2] Irshidat MR, Al-Saleh MH. Effect of using carbon nanotube modified epoxy on bond–slip behavior between concrete and FRP sheets. Construction and Building Materials. 2016;**105**:511-518

[3] Al-Zu'bi M, Fan M, Anguilano L. Advances in bonding agents for retrofitting concrete structures with fibre reinforced polymer materials: A review. Construction and Building Materials. 2022;**330**:127115

[4] Abdullah SR, Rosli FN, Ali N, Abd Hamid NA, Salleh N. Modified epoxy for fibre reinforced polymer strengthening of concrete structures. International Journal of Integrated Engineering. 2020;**12**(9):103-113

[5] Liu S, Chevali VS, Xu Z, Hui D, Wang H. A review of extending performance of epoxy resins using carbon nanomaterials. Composites Part B: Engineering. 2018;**136**:197-214

[6] Al-Zu'bi M, Fan M, Al Rjoub Y, Ashteyat A, Al-Kheetan MJ, Anguilano L. The effect of length and inclination of carbon fiber reinforced polymer laminates on shear capacity of near-surface mounted retrofitted reinforced concrete beams. Structural Concrete. 2021;**22**(6):3677-3691

[7] Al-Zu'bi M, Fan M, Anguilano L. Parametric investigation of flexural performance of concrete prisms retrofitted with near-surface mounted FRP bars. Composites Part C: Open Access. 2023;**12**:100421

[8] Johnsen BB, Kinloch AJ, Mohammed RD, Taylor AC, Sprenger S. Toughening mechanisms of nanoparticle-modified epoxy polymers. Polymer. 2007;**48**(2):530-541

[9] Quan D, Urdániz JL, Ivanković A. Enhancing mode-I and mode-II fracture toughness of epoxy and carbon fibre reinforced epoxy composites using multi-walled carbon nanotubes. Materials and Design. 2018;**143**:81-92

[10] Ashrafi B, Guan J, Mirjalili V, Zhang Y, Chun L, Hubert P, et al. Enhancement of mechanical performance of epoxy/carbon fiber laminate composites using single-walled carbon nanotubes. Composites Science and Technology. 2011;**71**(13):1569-1578

[11] Singh NP, Gupta VK, Singh AP. Graphene and carbon nanotube reinforced epoxy nanocomposites: A review. Polymer. 2019;**180**:121724

[12] Irshidat MR, Al-Saleh MH, Al-Shoubaki M. Using carbon nanotubes to improve strengthening efficiency of carbon fiber/epoxy composites confined RC columns. Composite Structures. 2015;**134**:523-532

[13] Rousakis TC, Kouravelou KB, Karachalios TK. Effects of carbon nanotube enrichment of epoxy resins on hybrid FRP–FR confinement of concrete. Composites Part B: Engineering. 2014;**57**:210-218

[14] Iwahori Y, Ishiwata S, Sumizawa T, Ishikawa T. Mechanical properties improvements in two-phase and three-phase composites using carbon nanofiber dispersed resin. Composites Part A: Applied Science and Manufacturing. 2005;**36**(10):1430-1439

[15] Zhou Y, Pervin F, Jeelani S. Effect vapor grown carbon nanofiber on thermal and mechanical properties of epoxy. Journal of Materials Science. 2007;**42**(17):7544-7553

[16] Al-Safy R, Al-Mahaidi R, Simon GP. Thermal and mechanical characterizations of nanomaterial-modified adhesive used in bonding CFRP to concrete. The Journal of Adhesion. 2011;**87**:842-857

[17] Zhu J, Wei S, Ryu J, Budhathoki M, Liang G, Guo Z. In situ stabilized carbon nanofiber (CNF) reinforced epoxy nanocomposites. Journal of Materials Chemistry. 2010;**20**(23):4937-4948

[18] Ladani RB, Wu S, Kinloch AJ, Ghorbani K, Zhang J, Mouritz AP, et al. Multifunctional properties of epoxy nanocomposites reinforced by aligned nanoscale carbon. Materials and Design. 2016;**94**:554-564

[19] Baruah P, Karak N. Bio-based tough hyperbranched epoxy/graphene oxide nanocomposite with enhanced biodegradability attribute. Polymer Degradation and Stability. 2016;**129**:26-33

[20] Bao C, Guo Y, Song L, Kan Y, Qian X, Hu Y. In situ preparation of functionalized graphene oxide/epoxy nanocomposites with effective reinforcements. Journal of Materials Chemistry. 2011;**21**(35):13290-13298

[21] Tk BS, Nair AB, Abraham BT, Beegum PS, Thachil ET. Microwave exfoliated reduced graphene oxide epoxy nanocomposites for high performance applications. Polymer. 2014;**55**(16):3614-3627

[22] Kumar A, Chouhan DK, Alegaonkar PS, Patro TU. Graphene-like nanocarbon: An effective nanofiller for improving the mechanical and thermal properties of polymer at low weight fractions. Composites Science and Technology. 2016;**127**:79-87

[23] Chandrasekaran S, Seidel C, Schulte K. Preparation and characterization of graphite nano-platelet (GNP)/epoxy nano-composite: Mechanical, electrical and thermal properties. European Polymer Journal. 2013;**49**(12):3878-3888

[24] Park S, Kim DS. Curing behavior and physical properties of an epoxy nanocomposite with amine-functionalized graphene nanoplatelets. Composite Interfaces. 2016;**23**(7):675-687

[25] Chen L, Zhao P, Xie H, Yu W. Thermal properties of epoxy resin-based thermal interfacial materials by filling Ag nanoparticle-decorated graphene nanosheets. Composites Science and Technology. 2016;**125**:17-21

[26] Thompson L, Nikzad M, Sbarski I, Yu A. Esterified cellulose nanocrystals for reinforced epoxy nanocomposites. Progress in Natural Science: Materials International. 2022;**32**(4):235-244

[27] Abraham E, Kam D, Nevo Y, Slattegard R, Rivkin A, Lapidot S, et al. Highly modified cellulose nanocrystals and formation of epoxy-nanocrystalline cellulose (CNC) nanocomposites. ACS Applied Materials & Interfaces. 2016;**8**(41):28086-28095

[28] Aziz T, Fan H, Zhang X, Khan FU, Fahad S, Ullah A. Adhesive properties of bio-based epoxy resin reinforced by cellulose nanocrystal additives. Journal of Polymer Engineering. 2020;**40**(4):314-320

[29] Peng SX, Shrestha S, Yoo Y, Youngblood JP. Enhanced dispersion and properties of a two-component epoxy nanocomposite using surface-modified

cellulose nanocrystals. Polymer. 2017;**112**:359-368

[30] Tanvir A, El-Gawady YH, Al-Maadeed M. Cellulose nanofibers to assist the release of healing agents in epoxy coatings. Progress in Organic Coatings. 2017;**112**:127-132

[31] Yue L, Maiorana A, Khelifa F, Patel A, Raquez JM, Bonnaud L, et al. Surface-modified cellulose nanocrystals for bio-based epoxy nanocomposites. Polymer. 2018;**134**:155-162

[32] Liu YL, Wei WL, Hsu KY, Ho WH. Thermal stability of epoxy-silica hybrid materials by thermogravimetric analysis. Thermochimica Acta. 2004;**412**:139-147

[33] Chisholm N, Mahfuz H, Rangari VK, Ashfaq A, Jeelani S. Fabrication and mechanical characterization of carbon/SiC-epoxy nanocomposites. Composite Structures. 2005;**67**(1):115-124

[34] Morshed SA, Young TJ, Chirdon WM, Zhang Q, Tatar J. Durability of wet lay-up FRP bonded to concrete with nanomodified epoxy adhesives. The Journal of Adhesion. 2018;**94**:742-767

[35] Hosur M, Mahdi TH, Islam ME, Jeelani S. Mechanical and viscoelastic properties of epoxy nanocomposites reinforced with carbon nanotubes, nanoclay, and binary nanoparticles. Journal of Reinforced Plastics and Composites. 2017;**36**(9):667-684

[36] Huttunen-Saarivirta E, Vaganov GV, Yudin VE, Vuorinen J. Characterization and corrosion protection properties of epoxy powder coatings containing nanoclays. Progress in Organic Coatings. 2013;**76**(4):757-767

[37] Jojibabu P, Zhang YX, Rider AN, Wang J, Prusty BG. Synergetic effects of carbon nanotubes and triblock copolymer on the lap shear strength of epoxy adhesive joints. Composites Part B: Engineering. 2019;**178**:107457

[38] Wang X, Tang F, Cao Q, Qi X, Pearson M, Li M, et al. Comparative study of three carbon additives: Carbon nanotubes, graphene, and fullerene-C60, for synthesizing enhanced polymer nanocomposites. Nanomaterials. 2020;**10**(5):838

[39] Montazeri A. The effect of functionalization on the viscoelastic behavior of multi-wall carbon nanotube/epoxy composites. Materials and Design. 2013;**45**:510-517

[40] Al-Zu'bi M, Anguilano L, Fan M. Effect of incorporating carbon- and silicon-based nanomaterials on the physico-chemical properties of a structural epoxy adhesive. Polymer Testing. 2023;**128**:108221

[41] Wang X, Xing W, Feng X, Yu B, Song L, Hu Y. Functionalization of graphene with grafted polyphosphamide for flame retardant epoxy composites: Synthesis, flammability and mechanism. Polymer Chemistry. 2014;**5**(4):1145-1154

[42] Nie Y, Hübert T. Effect of carbon nanofiber (CNF) silanization on the properties of CNF/epoxy nanocomposites. Polymer International. 2011;**60**(11):1574-1580

[43] Zhao S, Song Z, Cui J, Li C, Yan Y. Improving dispersion and integration of single-walled carbon nanotubes in epoxy composites by using a reactive noncovalent dispersant. Journal of Polymer Science Part A: Polymer Chemistry. 2012;**50**(21):4548-4556

[44] Guo J, Li Y, Wu S, Li W. The effects of γ-irradiation dose on chemical modification of multi-walled

carbon nanotubes. Nanotechnology. 2005;**16**(10):2385

[45] Kujur S, Pathak DD. Reduced graphene oxide-immobilized iron nanoparticles Fe(0)@rGO as heterogeneous catalyst for one-pot synthesis of series of propargylamines. Research on Chemical Intermediates. 2020;**46**(1):369-384

[46] Galvez P, de Armentia SL, Abenojar J, Martinez MA. Effect of moisture and temperature on thermal and mechanical properties of structural polyurethane adhesive joints. Composite Structures. 2020;**247**:112443

[47] Zhang Z, Zhang W, Li D, Sun Y, Wang Z, Hou C, et al. Mechanical and anticorrosive properties of graphene/epoxy resin composites coating prepared by in-situ method. International Journal of Molecular Sciences. 2015;**16**(1):2239-2251

[48] Shokrieh MM, Esmkhani M, Shokrieh Z, Zhao Z. Stiffness prediction of graphene nanoplatelet/epoxy nanocomposites by a combined molecular dynamic–micromechanics method. Computational Materials Science. 2014;**92**:444-450

[49] Sun LH, Ounaies Z, Gao XL, Whalen CA, Yang ZG. Preparation, characterization, and modeling of carbon nanofiber/epoxy nanocomposites. Journal of Nanomaterials. 2011;**2011**

[50] Zabihi O, Khayyam H, Fox BL, Naebe M. Enhanced thermal stability and lifetime of epoxy nanocomposites using covalently functionalized clay: Experimental and modelling. New Journal of Chemistry. 2015;**39**(3):2269-2278

[51] Sydlik SA, Lee JH, Walish JJ, Thomas EL, Swager TM. Epoxy functionalized multi-walled carbon nanotubes for improved adhesives. Carbon. 2013;**59**:109-120

[52] Thomas S, Rouxel D, Ponnamma D, editors. Spectroscopy of Polymer Nanocomposites. William Andrew: Norwich; 2016

[53] Mahmood N, Islam M, Hameed A, Saeed S. Polyamide 6/multiwalled carbon nanocomposites: Structure and properties. Materials Today. 2018;**14**:345-351

[54] Morimune S, Nishino T, Goto T. Graphene oxide reinforced poly(vinyl alcohol) nanocomposites. Composites Science and Technology. 2014;**93**:37-43

[55] Bhattacharyya A, Chen S, Zhu M. Graphene reinforced ultra-high molecular weight polyethylene with improved tensile strength and creep resistance properties. Express Polymer Letters. 2014;**8**(2):113-121

[56] Werner E, van Zyl M, Garcia M, Schrauwen BAG, Kooi GP, De Hosson JTM. Hybrid polyamide/silica nanocomposites: Synthesis and mechanical testing. Macromolecular Materials and Engineering. 2002;**287**:106-112

[57] Sadeghi M, Khanbabaei G, Dehaghani AHS, Sadeghi M, Aravand MA, Akbarzade M, et al. Gas permeation properties of ethylene vinyl acetate–silica nanocomposite membranes. Journal of Membrane Science. 2008;**322**(2):423-430

[58] Chen L, Zheng K, Tian X, Hu K, Wang R, Liu C, et al. Double glass transitions and interfacial immobilized layer in in-situ-synthesized poly(vinyl alcohol)/silica nanocomposites. Macromolecules. 2010;**43**(2):1076-1082

[59] Peng Z, Kong LX, Li SD, Spiridonov P. Poly(vinyl alcohol)/silica nanocomposites: Morphology and thermal degradation kinetics. Journal of Nanoscience and Nanotechnology. 2006;**6**(12):3934-3938

[60] Kumar A, Negi YS, Bhardwaj NK, Choudhary V. Synthesis and characterization of cellulose nanocrystals/PVA-based bionanocomposite. Advanced Materials Letters. 2013;**4**(8):626-631

[61] Alam MA, Samad UA, Anis A, Alam M, Ubaidullah M, Al-Zahrani SM. Effects of SiO_2 and ZnO nanoparticles on epoxy coatings and its performance investigation using thermal and nanoindentation technique. Polymers. 2021;**13**(9):1490

[62] Marathe U, Chalana A, Bijwe J. Performance augmentation of epoxy adhesives with TiN nanoparticles. ACS Omega. 2022;**7**(5):4150-4157

[63] Hong SK, Kim D, Lee S, Kim BW, Theilmann P, Park SH. Enhanced thermal and mechanical properties of carbon nanotube composites through the use of functionalized CNT-reactive polymer linkages and three-roll milling. Composites Part A: Applied Science and Manufacturing. 2015;**77**:142-146

[64] Ahn SN, Lee HJ, Kim BJ, Tan LS, Baek JB. Epoxy/amine-functionalized short-length vapor-grown carbon nanofiber composites. Journal of Polymer Science Part A: Polymer Chemistry. 2008;**46**(22):7473-7482

[65] Ma PC, Mo SY, Tang BZ, Kim JK. Dispersion, interfacial interaction, and re-agglomeration of functionalized carbon nanotubes in epoxy composites. Carbon. 2010;**48**(6):1824-1834

[66] Shengtao Z, Anyan G, Huanfang G, Xiangqian CH. Characterization of exfoliated graphite prepared with the method of secondary intervening. International Journal of Industrial Chemistry. 2011;**45**(6):2345-2350

[67] Pahlevanzadeh F, Emadi R, Setayeshmehr M, Kharaziha M, Poursamar SA. Antibacterial amorphous magnesium phosphate/graphene oxide for accelerating bone regeneration. Biomaterials Advances. 2022;**138**:212856

[68] Michels J, Cruz JS, Christen R, Czaderski C, Motavalli M. Mechanical performance of cold-curing epoxy adhesives after different mixing and curing procedures. Composites Part B: Engineering. 2016;**98**:434-443

[69] Lapique F, Redford K. Curing effects on viscosity and mechanical properties of a commercial epoxy resin adhesive. International Journal of Adhesion and Adhesives. 2002;**22**(4):337-346

[70] Morshed SA, Sinha A, Zhang Q, Tatar J. Hygrothermal conditioning of wet-layup CFRP-concrete adhesive joints modified with silane coupling agent and core-shell rubber nanoparticles. Construction and Building Materials. 2019;**227**:116531

[71] Morshed SA, Young TJ, Chirdon WM, Zhang Q, Tatar J. Durability of wet lay-up FRP bonded to concrete with nanomodified epoxy adhesives. The Journal of Adhesion. 2020;**96**(13):1141-1166

[72] Irshidat MR, Al-Saleh MH. Flexural strength recovery of heat-damaged RC beams using carbon nanotubes modified CFRP. Construction and Building Materials. 2017;**145**:474-482

[73] Al-Zu'bi M, Anguilano L, Fan M. Carbon-based and silicon-based nanomaterials for enhanced structural

adhesives. Solid State Phenomena. 2023;**354**:151-159

[74] Al-Zu'bi M, Fan M, Anguilano L. Near-surface mounted-FRP flexural retrofitting of concrete members using nanomaterial-modified epoxy adhesives. Journal of Building Engineering. 2024;**84**:108549

[75] Al-Zu'bi M, Fan M, Bertolesi E, Anguilano L. A review on retrofitting concrete members with near-surface mounted-fiber reinforced polymer composites. Structural Concrete. 2024;**25**(3):2242-2268

[76] Shhabat M, Al-Zu'bi M. A review of repairing heat-damaged RC beams using externally bonded- and near-surface mounted-CFRP composites. Composites Part C: Open Access. 2024;**354**:100519

Chapter 4

Isocyanate-Free, Room-Temperature Curing Polyurethanes for Structural Applications

Bruno Traber, Heiko Jung and Raphael Schaller

Abstract

This study presents the development and characterization of novel hybrid non-isocyanate polyurethanes (NIPUs) for structural adhesive applications. A one-pot synthesis method at room temperature was developed, combining polyfunctional cyclic carbonates, diethylenetriamine, and epoxy. Kinetic investigations revealed significant differences between bi- and tricyclic carbonates, with tricyclic variants demonstrating superior curing kinetics. The hybrid NIPUs exhibited remarkable lap-shear strengths of 14–16 MPa on untreated aluminum and planed beech wood after 12 hours of room-temperature curing. The absence of isocyanates offers health and safety advantages in various applications. The versatility of hybrid NIPUs was demonstrated through their broad adhesion spectrum and adaptable mechanical properties, positioning them as promising candidates for diverse industrial applications, including battery component encapsulation and multi-material laminates. While challenges remain, particularly in the availability of higher-functional cyclic carbonates, this research opens new avenues for high-performance adhesive systems.

Keywords: polyurethanes, non-isocyanate polyurethanes, room-temperature curing, structural, metal adhesion

1. Introduction

In 1937, Otto Bayer's pioneering discovery of polyurethanes laid the foundation for a polymer class now recognized for its exceptional versatility and applicability, as characterized by the urethane bond in its structure [1]. The diversity of polyurethanes is rooted in their modular synthesis, which allows the incorporation of a wide range of different repeating units. This adaptability enables fine-tuning of the material's physical and chemical properties, ranging from thermoplastics to highly reactive compounds. As a result, polyurethanes serve an expansive array of applications, including but not limited to foams, coatings, textile fibers, and adhesives [2–4].

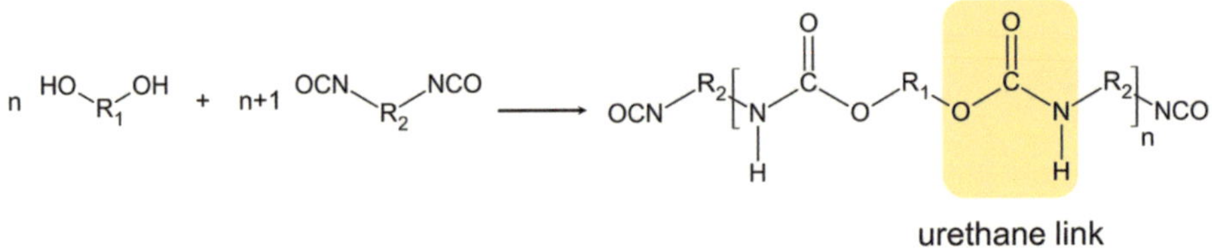

Figure 1.
Reaction equation of the polyaddition between polyol and isocyanate to form polyurethane.

The traditional synthesis of polyurethanes relies on a polyaddition reaction between polyols and isocyanates, typically employing at least bi-functional precursors (**Figure 1**). However, isocyanates pose health concerns, necessitating cautious use and obligatory training. Recent regulatory developments in the European Union [5] have catalyzed a quest for alternative synthetic pathways for polyurethanes. While conventional polyurethanes have dominated the literature with over a million publications in the past two decades, the emerging field of non-isocyanate polyurethanes (NIPUs) represents a niche, with approximately 4500 publications and 100 patents during the same period (**Figure 2**).

Within the research landscape, four primary approaches for synthesizing NIPUs have emerged [6, 7]:

1. Polyaddition of polyfunctional cyclic carbonates and amines: The advantage of this synthetic route is that with each urethane group, an additional functional group, a primary or secondary hydroxy group, is formed (**Figure 3**). Furthermore, no condensation products are formed during the polyaddition, which can be carried out at moderate temperatures (<120°C). An additional benefit of this synthesis is the biodegradability, potential bioavailability, and low toxicity of the cyclic carbonates.

2. Polycondensation between polycarbonates and polyamines or polycarbamates and polyols: The approach appears comparable to the previously described

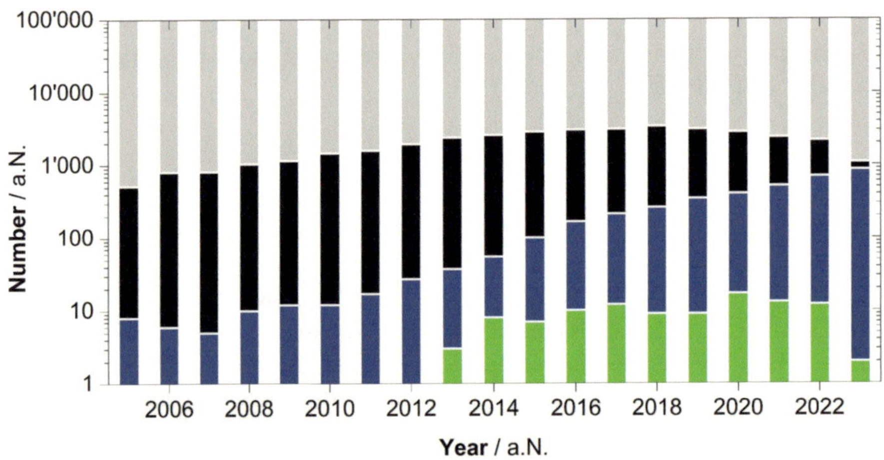

Figure 2.
Activities in filing patents and publishing articles on polyurethanes in the period from 2005 to 2023. Patents: ▪ isocyanate-free polyurethanes and ▪ polyurethanes. Articles: ▪ isocyanate-free polyurethanes and ▪ polyurethanes.

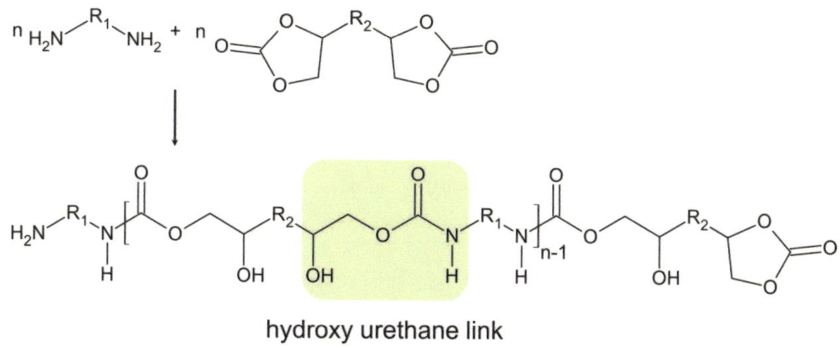

Figure 3.
Reaction equation of the polyaddition between amine and cyclic carbonate for the isocyanate-free formation of poly (hydroxyurethane).

polyaddition. These routes use polycarbonates and polyamines or polycarbamates and polyols with functionalities of at least two. **Figure 4** shows this is a polycondensation reaction where alcohol is produced as an undesired condensate, which is either distilled off or remains in the product.

3. Ring-opening polymerization of six- or seven-membered cyclic carbamates: This method initially appears to be elegant and free of by-products. However, it often requires very high temperatures, around 200°C, and catalysts such as sodium hydride to shift the equilibrium in the desired direction (**Figure 5**). Additionally, phosgene is typically used in the production of cyclic carbamates.

4. In situ formation of isocyanate groups, e.g., via Lossen, Hofmann, or Curtius rearrangement: In these complex synthetic pathways, the specified starting compounds undergo a rearrangement to form in situ isocyanates, which react with alcohol to produce urethanes (**Figure 6**). A disadvantage of this rearrangement is that the process, strictly speaking, is not isocyanate-free, as isocyanates are formed (albeit transiently) during the reaction.

Among these methods, the polyaddition of polyfunctional cyclic carbonates and amines proves to be particularly promising. It leads to the formation of

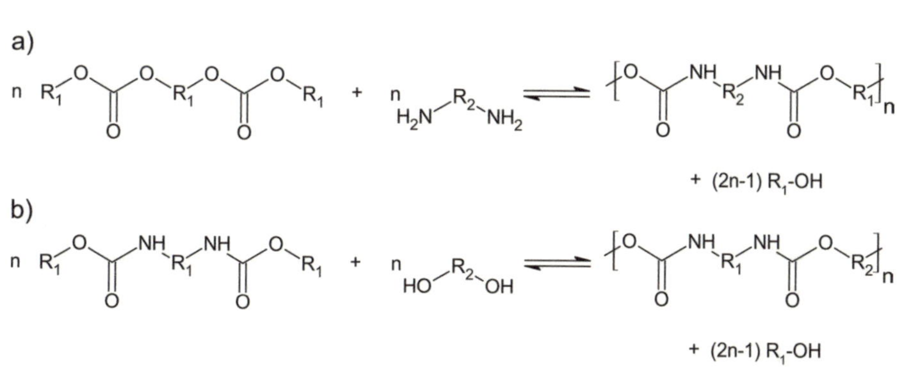

Figure 4.
These polycondensation routes employ (a) polycarbonates with polyamines or (b) polycarbamates with polyols to synthesize NIPUs.

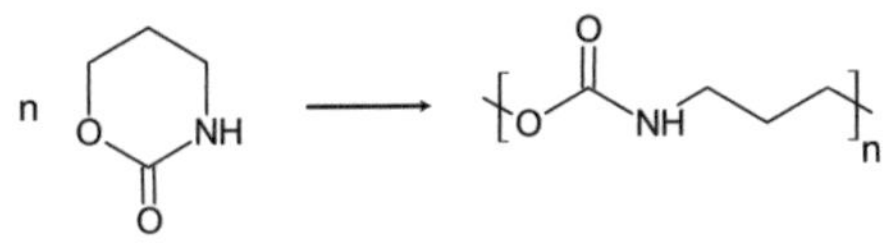

Figure 5.
Ring-opening polymerization of cyclic carbamates to NIPU.

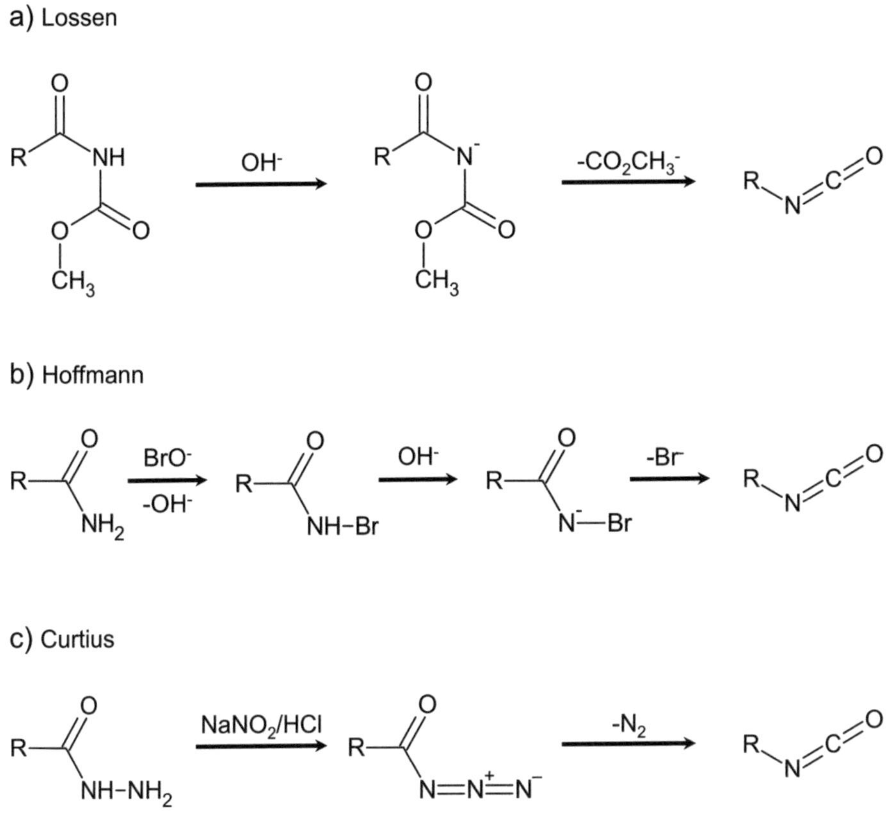

Figure 6.
(a) Lossen, (b) Hoffmann, or (c) Curtius rearrangement: In these complex synthetic routes, the specified starting compounds react through a rearrangement to form in situ generated isocyanates.

poly(hydroxyurethane) (PHU), which reveals urethane bonds and primary or secondary hydroxy groups (**Figure 3**). The motivation for this work lies not solely in the research and development of isocyanate-free polyurethanes but also in optimizing their properties for specific adhesive applications. In the course of this development, a promising combination of epoxides, polyfunctional cyclic carbonates, and amines was identified. These hybrid NIPUs can be tailored regarding their rigidity, flexibility, and adhesion spectrum, making them pre-destined for applications in adhesives, potting compounds, or impregnating resins. The scientific literature frequently describes multi-step processes for producing these hybrid NIPUs. These methods typically involve the initial synthesis of a prepolymer, followed by its further processing in a second step. Such processes are often complex and time-consuming, requiring multiple isolation and purification steps [8].

In an alternative approach to circumvent these multi-step procedures, Ke et al. developed a one-pot synthesis method [9]. In this method, all substances are combined simultaneously. The procedure involves mixing an epoxy-containing compound with a cyclic carbonate and an organocatalyst—in this case, 1,4-diazabicyclo[2.2.2] octane (DABCO)—followed by a reaction with an amine-containing compound. Although this method avoids multiple process steps, it requires heating the reaction mixture, which may limit its applicability in commercial settings such as carpentry workshops or construction sites.

In the following essay, a new generation of hybrid NIPUs are presented that can be produced and processed via a one-pot synthesis at room temperature within 1 hour. These NIPU adhesives are compared and evaluated in terms of viscosity range and lap-shear strength with other structural adhesive classes.

2. Investigations on reaction kinetics

In our efforts to optimize NIPU formulations, we conducted extensive kinetic studies to explore the interactions between formulation ingredients. We quantified the time-dependent viscosity evolution of various formulations using plate-plate rheometry from Anton Paar at 20°C, with a shear rate of 100 s^{-1} over a one-hour period. This approach allowed us to uncover the optimal synergy between polyfunctional cyclic carbonates and diethylenetriamine concerning their reaction kinetics.

Figure 7 provides a compelling visualization of the viscosity profiles for the NIPU formulations under investigation, contrasting the effects of different catalysts on bi- and tricyclic carbonates. Our analysis revealed significant differences in reaction kinetics: formulations with bicyclic carbonates exhibited a moderate viscosity increase from an initial range of 10–50 Pa·s to approximately 450 Pa·s within 1 hour.

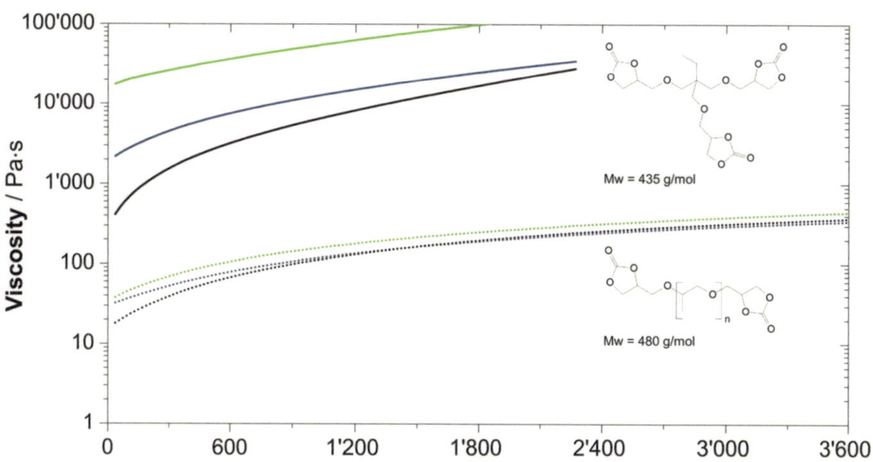

Figure 7.
Viscosity increase of NIPU formulations with polyfunctional cyclic carbonates and diethylenetriamine. A plate-plate rheometer was used at 20°C and a shear rate of 100 s^{-1}. Formulations with a bicyclic carbonate (poly(propylene glycol)α,ω-bis(cyclocarbonate), Mw = 480 g/mol): ... uncatalyzed, ... Catalyst 1, DBU, and ... Catalyst 2, TBD. Formulations with a tricyclic carbonate (trimethylpropane-tri(cyclocarbonate), Mw = 435 g/mol): — uncatalyzed, — Catalyst 1, DBU, and — Catalyst 2, TBD.

In contrast, those containing tricyclic carbonates demonstrated a dramatic rise to over 10,000 Pa·s in just 30 minutes. This pronounced increase in viscosity for tricyclic carbonates indicates enhanced curing kinetics at ambient temperature, highlighting their potential for rapid-curing NIPU systems.

The kinetic studies were conducted meticulously under controlled conditions, utilizing two distinct catalysts—1,5,7-triazabicyclo[4.4.0]dec-5-ene (TBD) and 1,8-diazabicyclo[5.4.0]undec-7-ene (DBU)—each at a 1% w/w concentration, while maintaining a stoichiometric ratio of 1:1 between carbonate and amine fractions.

This systematic approach not only facilitated a comprehensive evaluation of reaction kinetics but also laid the groundwork for developing optimized NIPU formulations with enhanced processing and performance characteristics. Future research may focus on further refining catalyst systems and fine-tuning the carbonate-amine stoichiometry to improve curing kinetics and achieve tailored properties for specific applications. Such advancements could potentially revolutionize the field of non-isocyanate polyurethanes, opening new avenues for their use in diverse industrial sectors.

3. Initial hybrid formulations

For preliminary experiments with hybrid NIPU formulations (**Table 1**), polyfunctional cyclic carbonates, diethylenetriamine, and epoxy were utilized.

The conversion of the carbonate with the amine to form C-N bonds was facilitated through targeted catalysis using organocatalysts. This choice of catalyst eased the

Formulation	**A**	**B**	**Remark**
Ingredients and composition	% w/w	% w/w	—
Epoxy	53.3	29.7	—
Cyclic carbonate	32.7	48.8	A: bi-, B: tricyclic carbonate
Catalyst	1.0	1.0	TBD catalyst
Amine	13.0	20.5	—
Properties	**A**	**B**	**Method**
Initial viscosity/Pa•s	6.0	324.0	DIN EN ISO 10364
Pot life at 20°C/min	5.0	6.0	DIN EN ISO 10364
Lap-shear strength/MPa A: planed beech wood B: untreated aluminum	6.0	9.8	A: DIN EN 12765 B: DIN EN 1465
Storage modulus at 20°C/MPa	1200.0	1200.0	ASTM D5279
Tensile strength/MPa	36.4	46.6	DIN EN ISO 527-1:2019
Strain at break/%	6.2	9.0	DIN EN ISO 527-1:2019
Glass transition temperature T_G/°C	51.6	49.6	ASTM D5279

Table 1.
Initial hybrid NIPU adhesive formulations A and B with their compositions and specific properties such as start viscosity, pot life at 20°C, lap-shear strength, storage modulus, glass transition temperature, and tensile properties.

desired crosslinking at room temperature, avoiding the need for curing at elevated temperatures (typically up to 120°C for several hours), which often leads to unwanted side reactions. Organocatalysts, such as tertiary amines and hydrogen-bond donors, exemplified by TBD or thiourea, effectively activate the carbonyl groups of cyclic carbonates at room temperatures [10]. The one-pot synthesis of the initial formulations followed a six-step protocol:

1. Homogenization of the epoxy and the polyfunctional cyclic carbonate using a speedmixer (2000 min^{-1} for 1 minute).

2. Integration of the catalyst followed by further homogenization (2500 min^{-1} for 3 minutes).

3. Thermal treatment to dissolve the catalyst in the epoxy-carbonate mixture (80°C for 40 minutes).

4. Cooling to room temperature and re-homogenization (2500 min^{-1} for 3 minutes).

5. Incorporation of diethylenetriamine at room temperature (2000 min^{-1} for 1 minute).

6. Immediate processing of the mass by casting or application for adhesive bonding at room temperature.

Detailed information regarding the synthesis methodology is disclosed in the patent application "Verfahren zur Herstellung eines hybriden nicht-isocyanathaltigen Polyurethans" by Lutz and Odermatt [11]. For industrial scaling, these hybrid NIPUs are designed as two-component adhesive systems, where Component A comprises the mixture of epoxy, cyclic carbonate, and catalyst. At the same time, Component B represents the amine component.

4. Isocyanate-free polyurethane hybrids for structural bonding

Based on preliminary findings, all subsequent formulations for hybrid NIPU adhesives with structural properties were engineered using the tricyclic carbonate trimethylpropane-tri(cyclocarbonate) with a molecular weight of 435 g/mol, selecting TBD as the catalyst due to its superior catalytic efficiency and an epoxy. **Figures 8** and **9** illustrate the application domains of various structural adhesives, presenting the relationship between lap-shear strength and initial viscosity. These two properties were measured as follows:

1. Lap-shear strength tests were performed on planed beech wood referred to DIN EN 12765 at durability class C1 (**Figure 8**) and untreated aluminum EN AW-5754 alloy according to DIN EN 1465 (**Figure 9**). All adhesives were prepared and mixed under controlled conditions of 20°C and 65% relative humidity prior to application. The substrates were used as received without pretreatment, with the following dimensions (length × width × thickness):

a. Beechwood: 80.0 mm × 20.0 mm × 5.0 mm, with a 10 mm overlap

b. Aluminum: 80.0 mm × 25.0 mm × 1.6 mm, with a 12.5 mm overlap

The lap-shear specimens were subjected to a pressure of 0.8 N/mm^2 for 12 hours at 20°C. Subsequently, the specimens underwent a 7-day conditioning period in a climate-controlled environment maintained at 20°C and 65% relative humidity. Lap-shear tests were performed using a ZwickRoell universal testing machine equipped with a 50 kN load cell and tested at a 10 mm/min crosshead speed.

2. Initial viscosity was measured using plate-plate rheometry from Anton Paar at 20°C with a shear rate of 100 s^{-1} according to DIN EN ISO 10364.

On beech wood, 2 K epoxy adhesives, atypical for wood joints, exhibit good lap-shear strengths of up to 13 MPa in the low viscosity range (200–4000 mPa•s).

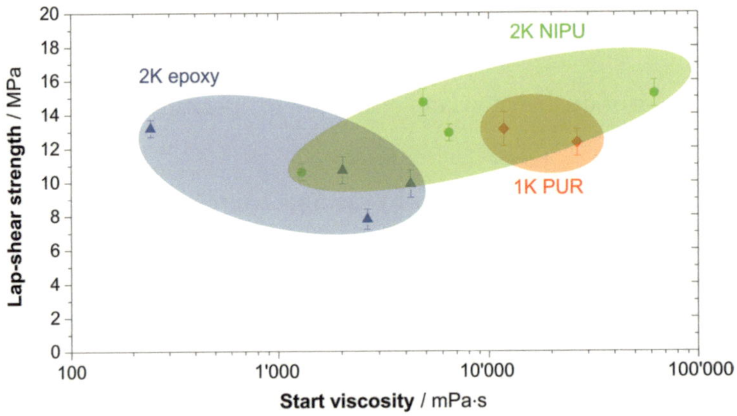

Figure 8.
Lap-shear strength on planed beech wood versus initial viscosity (measured after mixing) of structural adhesives: ▲ 2 K epoxy, ◆ 1 K PUR, and ● 2 K NIPU.

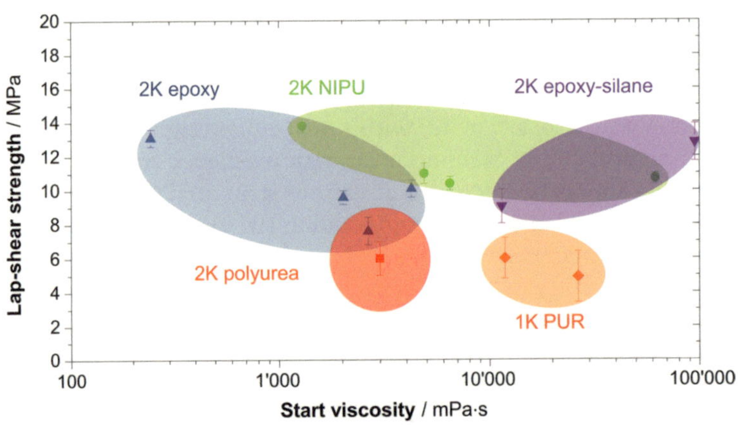

Figure 9.
Lap-shear strength on untreated aluminum EN AW-5754 versus initial viscosity of structural adhesives: ▲ 2 K epoxy, ■ 2 K polyurea, ◆ 1 K PUR, ▼ 2K epoxy-silane, and ● 2 K NIPU.

Reactive one-part polyurethane (1 K PUR) systems achieve lap-shear strengths of 12–13 MPa at typically higher initial viscosities >10,000 mPa•s. Hybrid NIPUs demonstrate the highest lap-shear strength of over 15 MPa across the broad viscosity range, with wood failure occurring in more than 60% of cases, indicating their versatility and effectiveness.

On untreated aluminum EN AW-5754, 1 K PUR and 2 K polyurea adhesives exhibit predictably poor adhesion due to the lack of surface pretreatment. In contrast, 2 K epoxy and 2 K epoxy-silane adhesives demonstrate superior performance, achieving lap-shear strengths ranging from 10 to 14 MPa.

Notably, hybrid NIPUs attain comparable results to 2 K epoxy and 2 K silane-modified adhesives on untreated aluminum, primarily exhibiting cohesive failure while maintaining their characteristic polyurethane tensile properties, as illustrated in **Figure 10**. This performance suggests that hybrid NIPUs may offer a promising alternative for bonding untreated aluminum surfaces without compromising the beneficial properties of polyurethane-based adhesives.

Stress-strain curves were generated in accordance with DIN EN ISO 527-1:2019. Epoxy and hybrid NIPU samples were prepared and mixed under controlled conditions of 20°C and 65% relative humidity before being injected into 5A tensile specimen molds. The specimens were then subjected to a 7-day conditioning period in a climate-controlled environment maintained at 20°C and 65% relative humidity. Tensile tests were conducted using an Instron universal testing machine equipped with a 30 kN load cell and tested at a 10 mm/min crosshead speed.

The stress-strain diagram (**Figure 10**) shows that by increasing the NIPU content in the epoxy, a transition from brittle to elastic behavior with strain-hardening behavior can be achieved. **Table 2** presents the corresponding lap-shear strengths. At an epoxy: NIPU ratio of 78:22, the highest shear strengths on the planed beech wood are achieved, with approximately 80% wood failure. On the aluminum alloy EN AW-5754, the maximum shear strength of 15.9 ± 1.0 MPa was obtained at an epoxy: NIPU ratio of 55:45, exhibiting cohesive failure in the adhesive. This result is remarkable for a polyurethane-based adhesive formulation on untreated aluminum.

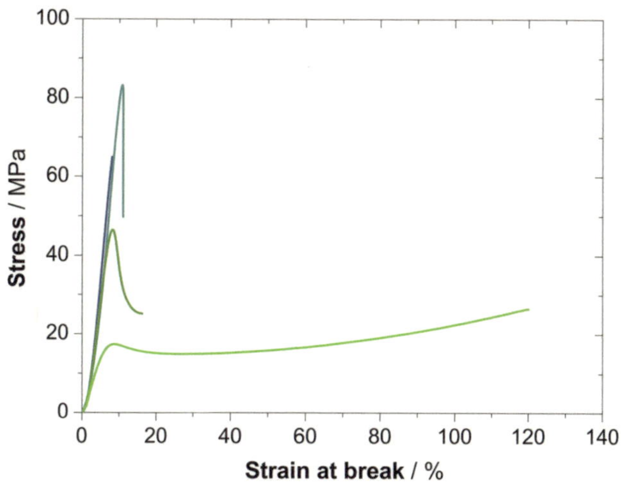

Figure 10.
The stress-strain diagram shows a transition from epoxy to a brittle and tough hybrid NIPU adhesive: Epoxy with – 0% w/w, – 22% w/w, – 33% w/w, and – 45% w/w NIPU content.

NIPU content in epoxy % w/w	Lap-shear strength planed beech wood MPa	Lap-shear strength untreated aluminum MPa
−00	11.3 ± 0.2	10.1 ± 0.4
−22	13.3 ± 0.4	10.8 ± 0.9
−33	12.9 ± 0.6	14.1 ± 0.5
−45	10.2 ± 0.2	15.9 ± 1.0

Table 2.
Lap-shear strength on planed beech wood and untreated aluminum EN AW-5754 of hybrid NIPU adhesives at various epoxy:NIPU proportions.

The adhesive demonstrates high toughness, with the epoxy contributing to adhesion on metals, supported by the additional hydroxy groups of the NIPU. This combination of properties opens new possibilities for developing high-performance, versatile adhesive systems for diverse structural applications.

5. Challenges and opportunities

The development and application of hybrid NIPU adhesives face several significant technical and economic challenges.

First, a fundamental issue lies in the generally low reactivity of cyclic carbonates, which impairs the efficiency of polyaddition reactions and leads to extended reaction times. This kinetic limitation can be problematic in industrial applications, negatively affecting processing speed and, consequently, process economics. Developing suitable catalyst systems to accelerate the reaction remains a central development goal.

Second, the current catalyst quantities are too high at 1% w/w. Therefore, optimizing catalyst systems to accelerate the polyaddition reaction is already a central research topic [12–14]. Advancements in this area could significantly increase processing speed and thus improve industrial applicability.

Third, the long-term stability of hybrid NIPU adhesives under application conditions has yet to be sufficiently researched. Comprehensive aging studies, including alternating climate change and weathering tests, are necessary to generate reliable data on the durability of bonded components.

Last, another challenge is the limited availability of higher-functional cyclic carbonates. While monofunctional cyclic carbonates are commercially available in large quantities, including from Huntsman or UBE Corporation, higher-functional variants are produced only on a laboratory scale, such as by Specific Polymers [15]. This limitation has significant implications for the scalability and economic feasibility of NIPU technology. Nevertheless, various approaches to producing cyclic carbonates from alternative sources are being explored in academia. Promising strategies include the use of polyether polyols from the fermentation of polysaccharides or the utilization of lignin or glycerine with subsequent cycloaddition of CO_2 as an abundant and cost-effective renewable raw material source to produce cyclic carbonates [6, 16–18]. Further activities also exist in areas where NIPU has carbonated soybean oil and a diamine comprising 100% renewable carbon content [19].

6. Conclusions

Hybrid NIPUs demonstrate their potential in structural bonding. In this context, initial formulations achieved remarkable lap-shear strengths of 14–16 MPa on untreated aluminum and planed beech wood, underscoring their suitability for demanding joints. The absence of isocyanates in these adhesive systems offers advantages in terms of health protection, particularly in carpentry workshops and construction sites, where exposure to potentially harmful substances can be minimized.

Of particular note are the surprising properties of hybrid NIPUs regarding their adhesion spectrum on untreated metal surfaces and their structural performance. Initial formulations achieved lap-shear strengths of nearly 16 MPa on untreated aluminum with a pressing time of 12 hours under room temperature conditions.

Further investigations indicated that new hybrid NIPU formulations with even higher lap-shear strengths in combination with increased toughness and reduced pressing times are achievable. This discovery opens new application fields for polyurethane adhesives in areas where pronounced metal adhesion is critical. A promising application area is the encapsulation of battery components in electromobility. Further, the expanded adhesion spectrum of hybrid NIPUs predestines them as laminating resins in multi-material laminates, such as those used in ski construction. This versatility in substrate compatibility and advantageous mechanical properties position hybrid NIPUs as a potential novel solution for complex bonding tasks across various industrial sectors.

Ultimately, the most significant limitation to the economic feasibility of this technology is the availability of higher-functional cyclic carbonates.

Acknowledgements

The authors thank Nicole O'Brien for her support with the text and Collano Technology and Business Development team for exciting and stimulating discussions on the topic, as well as Dr. Vincent Lutz and Daniela Odermatt for laying the technical foundation of isocyanate-free polyurethanes at Collano.

Author details

Bruno Traber, Heiko Jung and Raphael Schaller*
Collano AG, Neulandstrasse 1, Sempach Station, Switzerland

*Address all correspondence to: raphael.schaller@collano.com

References

[1] Bayer O. Das Di-Isocyanat-Polyadditionsverfahren (polyurethane). Angewandte Chemie. 1947;**59**:257-272. DOI: 10.1002/ange.19470590901

[2] Akindoyo JO, Beg MDH, Ghazali S, Islam MR, Jeyaratham N, Yuvaraj AR. Polyurethane types, synthesis and applications – A review. RSC Advances. 2016;**115**:114453-114482. DOI: 10.1039/C6RA14525F

[3] Market Data. Plastics Europe Regularly Publishes Reports on Market Data and on the Circularity of the European Plastics Industry. 2024. Available from: https://plasticseurope.org/resources/market-data/ [Accessed: December 11, 2024]

[4] Szycher M. Szycher's Handbook of Polyurethanes. 2nd ed. Boca Raton, Florida, USA: CRC Press; 2012. DOI: 10.1201/b12343

[5] Commission Regulation (EU). 2022. 2022/1149. 2022. Available from: https://eur-lex.europa.eu/legal-content/EN/TXT/PDF/?uri=CELEX:32020R1149 [Accessed: December 11, 2024]

[6] Błażek K, Datta J. Renewable natural resources as green alternative substrates to obtain bio-based non-isocyanate polyurethanes – Review. Critical Reviews in Environmental Science and Technology. 2019;**49**:173-211. DOI: 10.1080/10643389.2018.1537741

[7] Stachak P, Łukaszewska I, Hebda E, Pielichowski K. Recent advances in fabrication of non-isocyanate polyurethane-based composite materials. Materials. 2021;**14**:3497-3524. DOI: 10.3390/ma14133497

[8] Cornille A, Auvergne R, Figovsky O, Boutevin B, Caillol S. A perspective approach to sustainable routes for non-isocyanate polyurethanes. European Polymer Journal. 2017;**87**:535-552. DOI: 10.1016/j.eurpolymj.2016.11.027

[9] Ke J, Li X, Jiang S, Wang J, Knag M, Li Q, et al. Critical transformation of epoxy resin from brittleness to toughness by incorporating CO_2-sourced cyclic carbonate. Journal of CO_2 Utilization. 2018;**26**:30-313. DOI: 10.1016/j.jcou.2018.05.020

[10] Lutz V. Raumtemperaturhärtende NIPU-Systeme (non-isocyanate polyurethane). In: Proceedings Winterthurer Klebstofftagung. Winterthur: Zürcher Hochschule für angewandte Wissenschaften; 2023

[11] Lutz V, Odermatt D. Verfahren zur Herstellung eines hybriden nicht-isocyanathaltigen Polyurethans. Patent Application WO. 2024/165634 A1. 2024. pp. 27-32

[12] Carbajo-Gordillo AI, Benito E, Galbis E, Grosso R, Iglesias N, Valencia C, et al. Simultaneous formation of polyhydroxyurethanes and multicomponent semi-IPN hydrogels. Polymers. 2024;**16**:880-901. DOI: 10.3390/polym16070880

[13] Ochiai B, Kobayashi Y. Non-isocyanate synthesis of aliphatic polyurethane by $BiCl_3$-catalyzed transurethanization polycondensation. Polymers. 2024;**16**:1136-1150. DOI: 10.3390/polym16081136

[14] Kiełkiewicz D, Siewniak A, Gaida R, Greif M, Chrobik A. Ionic liquid catalysis in cyclic carbonate synthesis for the development of soybean oil-based non-isocyanate polyurethane

foams. Molecules. 2024;**29**:3908-3926. DOI: 10.3390/molecules29163908

[15] Poly(Ethylene Glycol), α,ω-Bis(Cyclocarbonate). 2024. Available from: https://specificpolymers.com/product/polyethylene-glycol-%ce%b1%cf%89-biscyclocarbonate/ [Accessed: December 11, 2024]

[16] Chen Q, Gao K, Peng C, Xie H, Zhao ZK, Bao M. Preparation of lignin/glycerol-based bis(cyclic carbonates) for the synthesis of polyurethanes. Green Chemistry. 2015;**9**:4546-4551. DOI: 10.1039/C5GC01340B

[17] Błażek K, Beneš H, Walterová Z, Abbrent S, Eceiza A, Calvo-Correas T, et al. Synthesis and structural characterization of bio-based bis(cyclic carbonate)s for the preparation of non-isocyanate polyurethanes. Polymer Chemistry. 2021;**12**:1643-1652. DOI: 10.1039/D0PY01576H

[18] Rollin P, Soares LK, Barcellos AM, Araujo DR, Lenardão EJ, Jacob RG, et al. Five-membered cyclic carbonates: Versatility for applications in organic synthesis, pharmaceutical, and materials sciences. Applied Sciences. 2021;**11**:5024-5080. DOI: 10.3390/app11115024

[19] Catalá J, Guerra I, García-Vargas JM, Ramos MJ, García MT, Rodríguez JF. Tailor-made bio-based non-isocyanate polyurethanes (NIPUs). Polymers. 2022;**15**:1589-1609. DOI: 10.3390/polym15061589

Section 3

Typical and Innovative Applications

Chapter 5

Adhesives: Use for Fixing Workpieces in Machining Operations

Nicolae Alin Panc

Abstract

The chapter analyzes the use of adhesives as an innovative and efficient method of part clamping for industrial machining, highlighting their advantages over conventional methods, as well as their limitations. In industry, fixing parts with complex shapes or low rigidity is a significant challenge, and traditional solutions do not always meet the requirements of precision and efficiency. The introduction of the adhesive clamping method has revolutionized the manufacturing process, offering a viable alternative. The chapter first explores the problem of clamping parts in fixtures, then introduces conventional and non-conventional methods of clamping. Emphasis is placed on presenting the two unconventional methods that utilize the properties of adhesives to fixturing the workpieces for machining. The advantages as well as limitations of the fixing method using adhesives are presented.

Keywords: clamping technology, fixtures, adhesive clamping, machining, non-conventional clamping

1. Introduction

The production of products is an attribute of human society without which it would not have evolved. Since ancient times, people have been searching for new ways of making products.

To make any kind of product, people need a technological system. This technological system can be simple or highly complex. By the term product we mean both an individual part, such as a simple object like a hammer, as well as a complex object made up of a large number of distinct parts that are assembled together to form subassemblies, assemblies, and finally the product itself, such as a car.

Whatever the type of product, it is necessary to realize each part individually. There are currently three ways of making a part: subtractive technologies, redistributive technologies, and additive technologies.

Subtractive technologies are those technologies in which a certain amount of material is removed from a workpiece to obtain the final shape of the part. This group of technologies includes turning, milling, broaching, grinding, and so on. These technologies are characterized by dimensional accuracy, high quality, and high cost price.

Redistributive technologies are those technologies that, starting from workpieces, redistribute its shape, either totally or only partially, to its final form. In the case of total redistribution, we are talking about molding technologies. Usually, this type of technology involves bringing the workpieces to a melting temperature. Partial redistribution technologies are represented by stamping, extrusion, forging, and so on. All these technologies have as their main attribute high productivity and low cost for high volume production, as well as lower quality of mechanical properties.

Additive technologies are those technologies that deposit material, layer by layer, until the desired part shape materializes. These technologies are represented by fused deposition modeling (FDM), selective laser sintering (SLS), solid laser melting (SLM), stereolithography (SLA), and so on. These technologies have as their main attribute the simplicity of the technology, the relatively low cost, as well as the low quality of the product obtained.

Whichever manufacturing technology is chosen, the basic elements of the manufacturing system are: machine tool, tools, fixtures, and workpiece. The machine tool performs the main movements and generates the work parameters, both energy and information. Tools act on the workpieces to bring them to the desired shape. Workpiece is the material from which the part is made and which, under the action of the tools, undergoes shape transformations. Fixtures mean the elements that hold the workpiece in a unique position during the time the tool acts on the workpiece. Fixtures may also act to facilitate certain movements of the workpiece or tool.

There are a variety of fixtures, which have specific purposes, and their choice depends on the type of technology used, the shape of the workpieces, the forces involved in the working process, and so on.

Fixtures can be categorized according to a number of criteria:

a. The clamping principle: conventional clamping and non-conventional clamping.

b. Degree of universality: universal, special, specialized, modular.

c. Type of actuating force: mechanical, pneumatic, hydraulic, magnetic.

d. By operation type: manual, mechanized, automated.

e. The role of fixtures: fixing, assembly, measuring, and so on.

In principle, for technical and economic reasons, there are five basic rules when determining the type of clamping and the fixture used.

The first rule - the fixing method should be as simple as possible.

Second rule - the cost of fixing should be as low as possible.

Third rule - the fixture should ensure safety during the machining process.

Fourth rule - clamping should not influence the quality of machining.

Fifth rule - clamping time should be as short as possible.

By following these rules and taking into account the volume of production, different clamping methods should be chosen on a case-by-case basis. Thus, in the case of a small production series, we will look for fastening methods that require the simplest possible fixtures, and the most universal fastening solutions will be used, even if the fastening time will be long. In the case of large production series, we will always choose those fixtures that allow the shortest possible fastening times in order to ensure high productivity, and so, we turn to special fixtures.

The fixture and the clamping principle has an important role to ensure that it has a direct influence on the machining accuracy, product quality, and cost price. Thus, understanding and finding the most suitable clamping methods play an important role in determining the technologies to obtain products. While until recently conventional clamping fixtures were predominantly used, a number of non-conventional clamping fixtures have recently emerged that come with a number of advantages. However, this would not have been possible without research, on the one hand, and industry needs, on the other hand, which could no longer be met by conventional clamping fixtures.

Research over the last 30 years has led to the discovery of new materials with properties superior to those known, particularly in the field of adhesives and their applications.

The complexity of the products has continually evolved, and solutions have had to be found for the processing of increasingly complex shaped parts that are often thin-walled and have low rigidity. Conventional clamping fixtures were simply inadequate for the new requirements, so new ways of clamping were needed. In the remainder of this chapter, the main methods of non-conventional clamping with their advantages and limitations will be presented. It will also present the working principle of these clamping methods and how they can be used in industrial practice and possible research directions in this field.

2. The principles of fixing

Fixing the workpieces in the fixture means, in a simplistic way, the elimination of the degrees of freedom of the workpiece so that its position is unambiguously maintained throughout the machining process. However, in addition to the elimination of degrees of freedom, the fixture must also ensure the positioning of the workpieces. Positioning means that each time the workpieces is placed in the fixture, it must end up in exactly the same position as before it.

2.1 Positioning

If the workpieces are positioned on the machine tool, the machine tool does not have to be adjusted each time because the adjustments are retained from the first workpiece. Basically, positioning reduces work time by eliminating the set-up time for a batch of workpieces. Therefore, the possibility of the fixture to position the workpieces is absolutely necessary when machining a batch of workpieces of the same type. Fixtures that do not perform positioning will not be encountered unless individual workpieces are being processed.

The most commonly used positioning principle is called the 3-2-1 principle, which refers to the fact that the workpiece is enclosed between three surfaces called bases that cancel 3, 2 and 1 degrees of freedom, respectively, as follows (**Figure 1**):

- The surface that cancels three degrees of freedom is called the placement base.
- The surface that cancels two degrees of freedom is called the guidance base.
- The surface that cancels a degree of freedom is called the stop base.

The 3-2-1 principle has the advantage of simplicity except that it limits the number of surfaces that can be machined on a workpiece. Thus, of the six surfaces of a workpieces, which comprise the workpiece in a parallelepiped, only three surfaces can be machined integrally. But after positioning, the workpieces are to be clamped, which means that one more surface of the workpiece cannot be machined. The result of positioning and clamping is that only two surfaces can be machined completely (**Figure 1(a)**).

In the case of workpieces with transverse holes, the pin-positioning principle can also be used (**Figure 2**). This positioning principle still uses three surfaces for the base, of which one is a flat surface, called the placement base, and two circular surfaces. The advantages of this positioning principle are that a maximum of five surfaces can now be machined, but this principle can only be used on workpieces that have been initially machined and have holes in them that can be used for positioning.

Cylindrical workpieces can be positioned using either the 3-2-1 principle or the pin positioning principle. In all cases, it is advisable to avoid large-area contact between the support elements and the surfaces of the workpieces, as clamping errors occur. For this reason, a linear contact exists between the workpieces and the positioning elements.

2.2 Fixing

Once the positioning is achieved, the total degrees of freedom are completely eliminated by using different types of clamping mechanisms that develop the clamping force Fc. The clamping force is applied by the clamping mechanism to the workpieces and is taken up by the positioning elements, usually the placement base or the guidance base, as these are the elements that have the maximum rigidity.

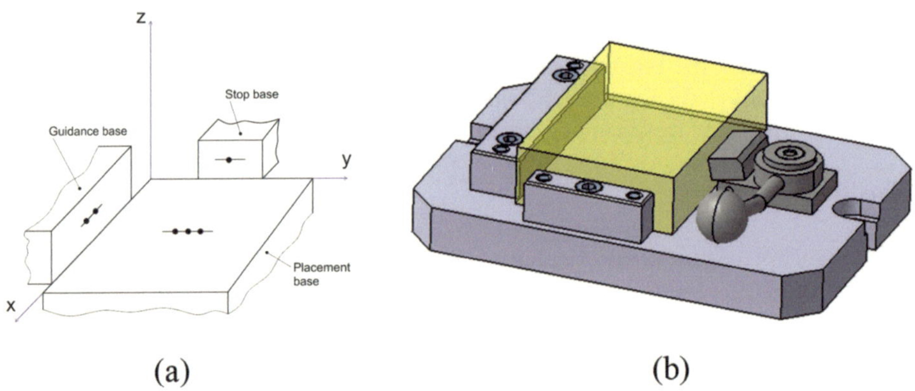

Figure 1.
Principle of positioning 3-2-1: (a) the basics, (b) fixture example.

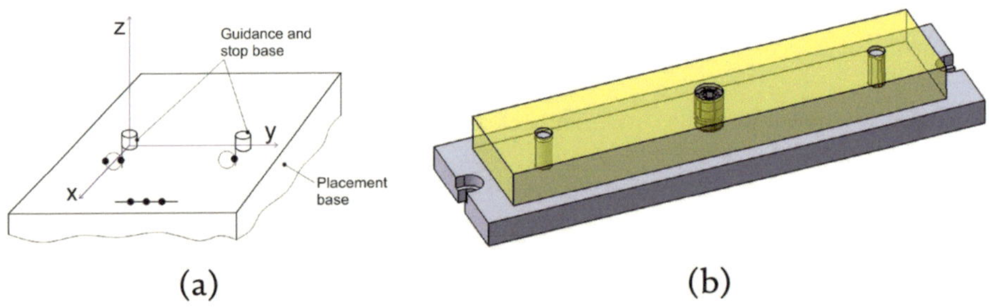

Figure 2.
Principle of positioning on pins: (a) the basics, (b) fixture example.

DOI: http://dx.doi.org/10.5772/intechopen.1008769

The clamping force shall be determined taking into account the relation 1:

$$R(F_a, F_n) < F_c < F_{10} \tag{1}$$

where:

$R(F_a, F_n)$ - represents the resultant of the forces acting on the workpiece or part and tend to move it during processing: the cutting force (F_a), other forces that appear during the processing (F_n), in [N].

F_c- Clamping force, in [N].

F_{10}- Maximum force to reach the proportionality limit σ_{10}.

After fixing, the workpieces can be in one of three situations (**Figure 3**):

- Pushed to placement base (a).
- Pull to placement base (b).
- Push to guidance base (c).

To ensure good productivity, it would be advisable to have a minimum number of workpieces clamped in the fixture, that is, it should be possible to process a maximum

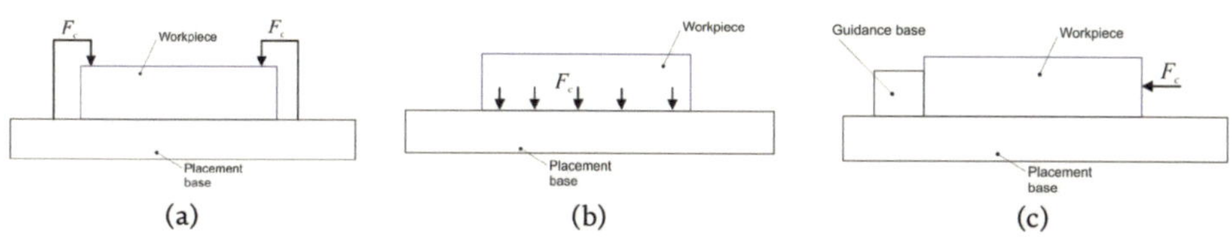

Figure 3.
Types of clamping in the fixture.

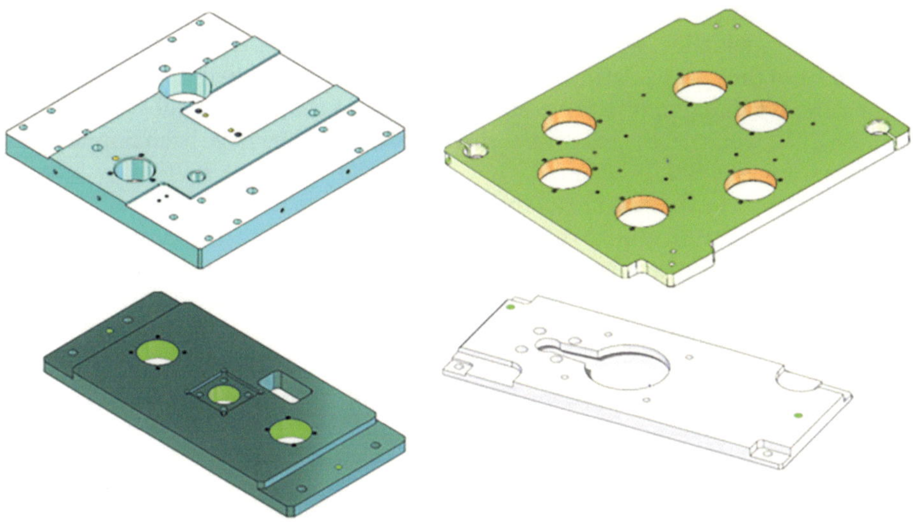

Figure 4.
Types of plate parts.

number of surfaces in one clamping. This desideratum can be achieved if a fixture is applied whereby the workpieces are pulled onto the machine table. However, this type of clamping can only be achieved in the classical way by using magnetic or vacuum clamping. Unfortunately, both these methods of clamping are strongly conditional as follows:

a. magnetic clamping is only possible for workpieces with ferromagnetic properties;

b. vacuum clamping only for workpieces without through-holes;

c. for both variants, the contact surface between the workpieces and the placement surface of the fixture should be as large as possible.

Therefore, clamping by pulling the workpieces on the table is only possible for a type of workpieces that have a large surface area, such as those shown in **Figure 4**.

3. Techniques of adhesive fixing

Due to the complexity of new products, there are a number of parts in new products that have low rigidity or complex shapes. For the machining of such parts, the conventional positioning and clamping methods presented above are often difficult to use.

In the case of such parts, non-conventional clamping often remains the optimal solution. In the literature, only Joshi in [1] refers to this method in a succinct way, while the rest of the specialized works such as Ref. [2–5] only mention them. A number of scientific research present these methods and study their effectiveness. Thus, Raffles in Ref. [6] outlines some of the most widely used non-conventional clamping along with their advantages and limitations.

There are two methods in the category of non-conventional clamping methods, namely: encapsulation clamping and adhesive clamping.

The encapsulation method consists of embedding workpieces in different materials such as low-melt alloy, wax, or water ice. After processing, the workpieces are extracted from the encapsulating material. This method is studied in Ref. [7, 8].

The adhesive fixing method uses adhesives to bond the part to a substrate. After machining, the workpieces peel off the surface to which they have been bonded. This method is studied in Ref. [9–12].

Both methods use the same principle, i.e., utilizing the retention capacity of the workpieces in the fixture due to the coefficient of friction. It occurs between the workpieces and the fixture placement surface composed of the encapsulating material or the adhesive used, as appropriate. In addition to this basic principle, each method has its differences, namely:

- In the case of encapsulation, mechanical restraint also occurs if the material used for encapsulation fixing can penetrate into the inner surfaces of the contour boundary (**Figure 5a**).

- In adhesive fixing method, chemical interaction between the adhesive and the surface layer of the workpieces may also occur (**Figure 5a**).

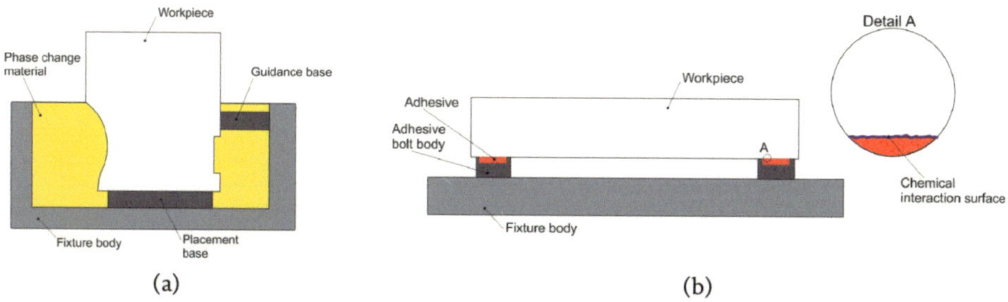

Figure 5.
Possible deviations from the method of fixation.

De Meter presents in Ref. [9] a method of clamping using adhesive bolts that he later transforms into a commercial product, which have as main advantages: maximum workpiece accessibility, flexibility of holding position, high holding strength, and short cycle times.

Panc in Ref. [11] comes up with another approach that uses, instead of adhesive bolts, adhesive tapes. His method has as advantages the simplicity of the method, minimization of the chatter phenomenon, and possibility to fix parts with low stiffness. However, the method proposed in Ref. [11] has the disadvantage of lower holding capacity than the method proposed by De Meter in Ref. [9].

3.1 Clamping using adhesive bolts

Fixing on adhesive bolts is the most widely used method today. The working principle of this method is simple and consists of the following steps:

Step 1: Applying adhesive to the adhesive bolts.

Step 2: Placing the workpieces on the bolts.

Step 3: Waiting for a while for the adhesive to bond or activate.

Step 4: Machining the workpieces.

Step 5: Removing the adhesive bond: mechanically, thermally, or chemically.

Step 6: Cleaning the contact surface between the workpieces and the adhesive bolts.

Figure 6 shows an example of a fixture using four adhesive bolts to clamp a workpiece.

As presented in Ref. [11], the use of this clamping method is viable if the following aspects are considered:

a. The workpiece must be rigid—so that no elastic deformations occur during machining due to cutting forces. Also, if the rigidity is low, chatter may occur, and deterioration of the machined surface quality may occur.

b. The adhesive used must bond effectively, ensuring a solid bond—depending on the type of material, different adhesives are used to achieve the highest possible holding strength.

c. The surface on which the bonding has been made will be further machined—as often with this method the bonding is to a workpiece surface that is then machined. The principle is shown in **Figure 7**.

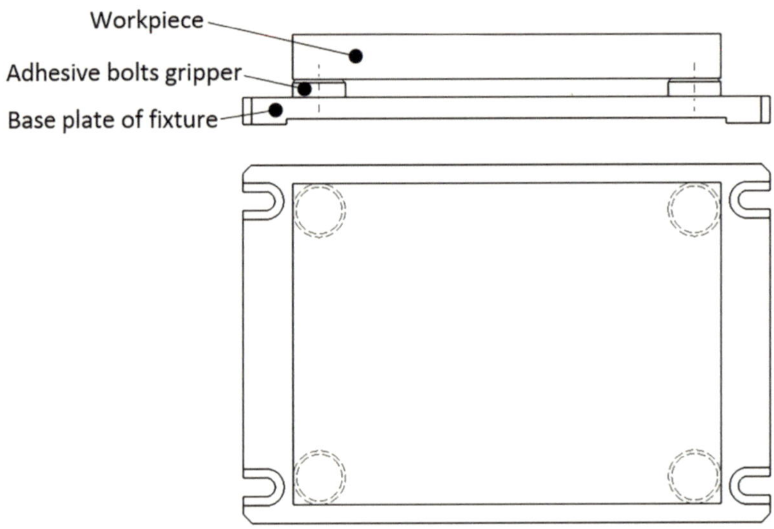

Figure 6.
Fixing a part by adhesive bolt gripper.

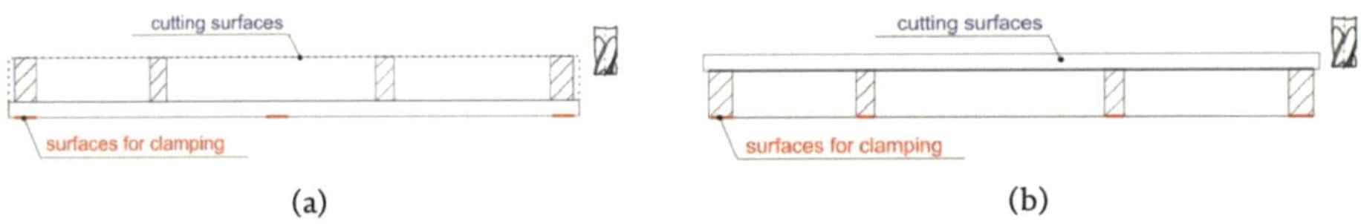

Figure 7.
Steps in machining a part with low stiffness: (a) first clamping-in, (b) second clamping-in.

For fastening using adhesive bolts, the second fastening step is carried out on the already machined surface. This is not an obstacle if the de-bonding of the adhesive is carried out by thermal methods; there is a solvent that can dissolve the adhesive, or the bonding surfaces can be cleaned without leaving any traces on the part. The temperature de-bonding method is only suitable for metal parts; applied to plastics, this method can damage the part.

d. During the de-bonding process, the part must not deform, that is, those de-bonding methods that do not introduce deformation into the part will be used. If this is not relevant for rigid parts, it is difficult or even impossible for low stiffness parts.

e. The costs associated with this method of clamping must be low, that is, the cost of clamping must be economically justified, otherwise classical clamping variants such as vacuum clamping or magnetic clamping can be used.

f. The time required for bonding and de-bonding should be short—while this is not usually taken into account when processing a single workpiece, it is important to keep these times as short as possible when processing a batch of workpieces.

g. Adhesive fixing systems to be truly reliable.

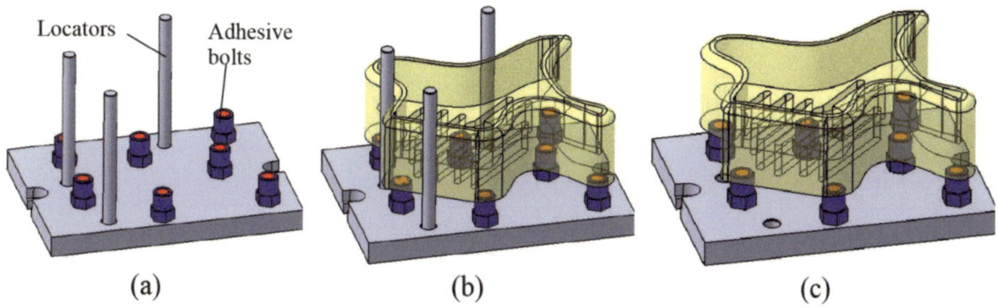

Figure 8.
Positioning and fixing workpieces using adhesive bolts.

Fixtures of this type are simple constructions that consist of a base plate that is mounted on the machine tool table, a minimum of four adhesive bolts that are fixed to this base plate, and auxiliary positioning elements (locators) that can be removed at a later stage to facilitate contour machining. The auxiliary elements are used when machining a batch of workpieces. **Figure 8** shows an example of how to work with such a fixture, which is first equipped with the adhesive bolts and auxiliary positioning elements (a). In the next step, the workpieces are positioned and fixed to the adhesive bolts (b), and after fixing, the auxiliary positioning elements can be removed from the fixture (c). In this way, the workpieces can also be machined on the contour and not only on the top surface. For the next workpiece to be machined, first reposition the locators in the base plate. The second workpiece will then be positioned identically to the first one, and the work method described above is repeated.

There are two types of adhesives used to fix the workpieces to the work bolts, namely, usual and special adhesives.

Usual adhesives are adhesives that are commonly used in industry, due to their low cost and efficiency. This category of adhesives includes polyacrylate-based adhesives, latex-based adhesives, and one-component epoxy-based adhesives. The problem in the use of these adhesives is related to the time needed to achieve the adhesive bond, which ideally should be as short as possible.

Special adhesives are adhesives where the bonding process is activated either chemically or physically. Chemically activated adhesives include two-component adhesives such as epoxy and polyurethane resins. These form strong bonds when activated. In the category of physically activated adhesives are photoactive adhesives, which by exposure to ultraviolet radiation initiate the polymerization reaction, leading to an active and controlled curing.

While there is no relevant scientific research on the use of universal adhesives, there is relevant scientific research on the use of specialty adhesives, which provides valuable information such as that presented in Ref. [6, 9, 10, 12, 13].

3.2 Clamping using adhesive tape

The adhesive tape clamping method is an alternative to clamping using adhesive bolts and overcomes the main drawback of this method, which is that it cannot clamp parts with low rigidity.

This method consists of bonding the workpieces over the entire placement surface using adhesive tape. As a result, the contact pressure is distributed evenly over the entire contact surface of the workpieces, which provides superior stability. Although

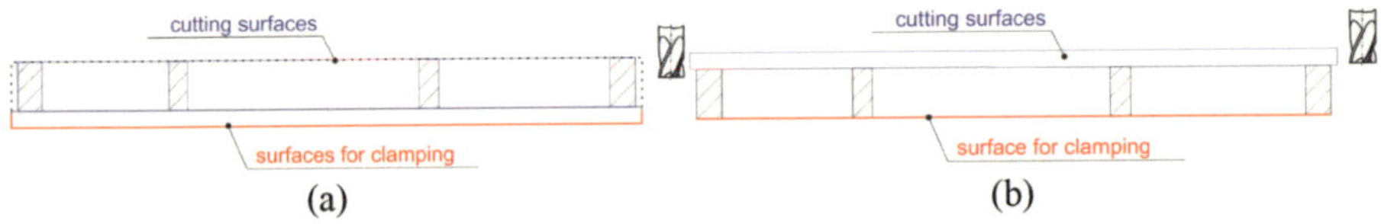

Figure 9.
Steps in machining a part with low stiffness: (a) first clamping-in, (b) second clamping-in.

the adhesive holding force may be lower than with adhesive bolts, its action over a large area generates an acceptable holding force. In addition, clamping on a large base increases part rigidity and reduces vibration during machining, which is a substantial advantage. The fact that the fixture is very simple, using only one base plate, and the low cost of adhesive tape are also significant advantages.

Figure 9 shows the principle of using this method in the case of a workpiece with low stiffness, which would not be possible to fix in any other way. In this case, too, a technological addition is used, which allows machining of the part in the first clamping-in (**Figure 9(a)**), and in the second clamping-in, only the technological addition is machined (**Figure 9(b)**). However, there is also the possibility of bonding the workpiece brought to the final thickness to a plate acting as a placement surface, and when machining the workpieces, the tool also enters the placement surface. This eliminates the use of technological additives but sacrifices the placement surface.

With all the advantages it brings, there may be some drawbacks in using this method, namely:

- the effective holding capability of the part using adhesive tape.
- the possibility of obtaining parts with a dimensional accuracy in the IT6-IT7 accuracy class according to ISO 286.
- possibility to fix as many types of materials as possible.
- resistance of the bonding to a dynamic regime specific to the machining process and to the action of the cutting fluid, which can negatively influence the adhesive bond's holding capacity.

The holding capacity of the tape is influenced by many factors, the most important of which are the type of adhesive used and the roughness of the surfaces to which the

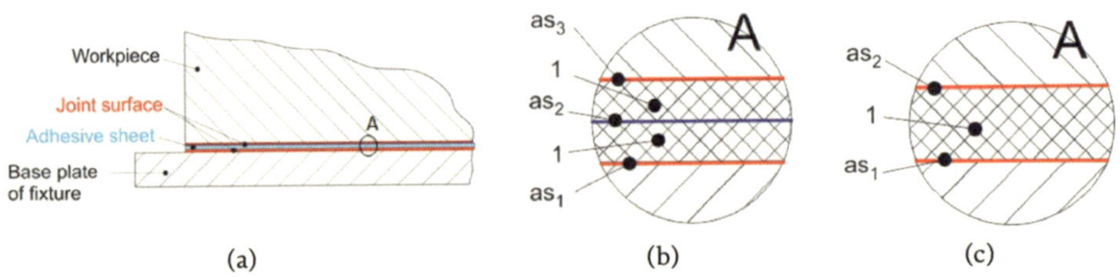

Figure 10.
Interface workpiece-adhesive-fixtures: (a) setup [11], (b) using single-sided adhesive tapes, (c) using thin double-sided tape.

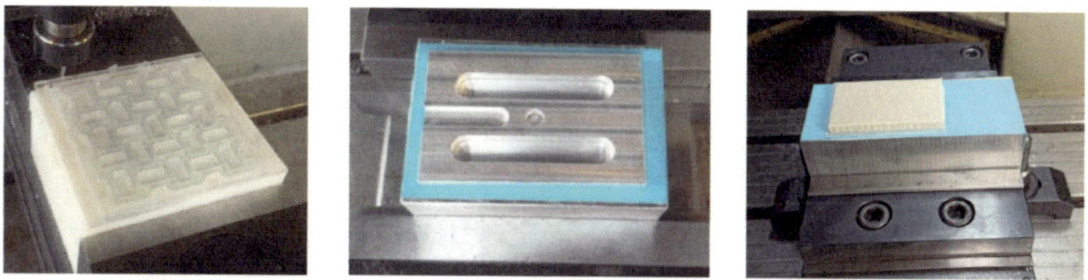

(a) (b) (c)

Figure 11.
Examples of adhesive fixing for machining for: (a) POM-C, (b) AW 5083, (c) PEEK.

tape is applied. Adhesive tape can be single-sided adhesive tapes and a thin double-sided tape. **Figure 10** shows the workpiece-adhesive-fixtures interface.

When single-sided adhesive tapes are used, two 1 tapes are used, see **Figure 10**, one is bonded to the workpiese and the other to the base plate of fixture. The two tapes are bonded together with a cyano-acetylated universal adhesive. Basically, three adhesive bonding surfaces are formed: as_1, as_2, and as_3, and as_2 has the highest holding capacity.

When thin double-sided tape is used, a single bend 1 is used, which forms two adhesive bonding surfaces: as_1, as_2.

From the studies carried out in Ref. [11], it has been established that the dimensional accuracy of the parts is realized in the accuracy class IT7-IT6, and the dimensions are in accordance with ISO 287.

Following the processing on different materials such as aluminum alloy (AW 5083), steel, polyoxymethylene (POM-C), and Poly-Ether-Ether-Ketone (PEEK), the results were optimal. The method is viable for a wide range of metallic or plastic materials.

Figure 11 shows three examples of the use of adhesive tape with using thin double-sided tape (a) and single-sided adhesive tapes (b) and (c) for processing parts with low rigidity. For each type of material, the clamping force was high enough to mill workpieces.

4. Advantages and limitations of the adhesive clamping method

Like any technology available today, adhesive clamping has its advantages and limitations. Although a number of advantages have been mentioned above, it is now time to present them in a summarized way.

a. Advantages of the method

- The method is suitable for parts with complex shape or low rigidity.
- The fixture used has a simple shape, and the cost of making it is minimal.
- It does not require high-level specialized knowledge to use and operate this method.
- It can effectively eliminate the chatter phenomenon that occurs when machining low stiffness blanks.

- The costs of using the method is minimal, which makes it particularly recommendable when machining single parts or small series.

- It can be combined with the vacuum method as wel as with other conventional methods to increase clamping force.

b. Limitations of the method

- There is always a risk of accidental detachment of the workpiece if the adhesive bond fails. While in the case of conventional clamping, safety measures can be taken, in this case, no such measures can be taken, otherwise some of the advantages of this method would be lost.

- The surface of the workpiece and fixture must be carefully cleaned, which may increase clamping time.

- For each type of material, preliminary investigations must be carried out to be sure that the adhesive used produces a bond as expected. This shortcoming can be eliminated by using specialized studies, but these can be expensive or time-consuming.

- There is a possibility that the adhesive may carry out chemical reactions that can degrade the surface of the workpiece, and this can be a drawback in certain situations, especially for parts that have an esthetic role in a product.

- There is no universal adhesive that is suitable for bonding all materials, so different types of adhesives are needed.

5. Conclusions

As presented in this chapter, adhesives also have an important role to play in the field of fixturing workpieces for machining. Thus, according to the scientific research mentioned above, adhesive clamping is a viable clamping method with outstanding results for all types of parts. For intense machining regimes, it is recommended to use clamping using adhesive bolts, which has been tested for both milling and grinding operations and has proven to be viable and competitive with classical clamping methods. For moderate machining regimes, and in particular for workpieces with low rigidity, clamping using adhesive tape has also proven viable and has been tested in milling operations.

It should be recalled that adhesive clamping is a non-conventional clamping method and is complementary to conventional clamping methods. Always consider both the cost and time required for clamping when determining the clamping method. Only in this way can a correct decision be made as to which type of clamping method to use. For safety reasons, however, this method is preferred only for semi-finished products of high complexity or low rigidity, otherwise conventional clamping is preferable.

Author details

Nicolae Alin Panc
Technical University of Cluj-Napoca, Cluj-Napoca, Romania

*Address all correspondence to: nicolae.panc@tcm.utcluj.ro

References

[1] Joshi PH. Jigs and Fixtures: Design Manual. 2nd ed. New York: McGraw-Hill; 2003. 237 p. McGraw-Hill Professional Engineering

[2] Nee AYC, Whybrew K, Senthil Kumar A. Advanced Fixture Design for FMS [Internet]. London: Springer London; 1995. DOI: 10.1007/978-1-4471-2117-6 [Accessed: June 17, 2022]

[3] Qin G. Advanced Fixture Design Method and its Application [Internet]. Singapore: Springer Singapore; 2021. DOI: 10.1007/978-981-33-4493-8 [Accessed: June 17, 2022]

[4] Venkataraman K. Introduction to jigs and fixtures. In: Design of Jigs, Fixtures and Press Tools [Internet]. Cham: Springer International Publishing; 2022. pp. 3-8. DOI: 10.1007/978-3-030-76533-0_1 [Accessed: June 17, 2022]

[5] Hoffman EG. Jig and Fixture Design. 5th ed. Clifton Park, NY: Thomson/Delmar Learning; 2004. 369 p

[6] Raffles MH, Kolluru K, Axinte D, Llewellyn-Powell H. Assessment of adhesive fixture system under static and dynamic loading conditions. Proceedings of the Institution of Mechanical Engineers, Part B: Journal of Engineering Manufacture. 2013;**227**(2):267-280

[7] Sarma SE, Wright PK. Reference free part encapsulation: A new universal fixturing concept. Journal of Manufacturing Systems. 1997;**16**(1):35-47

[8] Lee E, Sarma SE. Reference free part encapsulation: Materials, machines and methods. Journal of Manufacturing Systems. 2007;**26**(1):22-36

[9] De Meter EC. Light activated adhesive gripper (LAAG) workholding technology and process. Journal of Manufacturing Processes. 2004;**6**(2):201-214

[10] De Meter E, Chow KH, Marsh E. Methodology of using PAAW and the underlying support network of an L-PBF part to facilitate machining. Procedia Manufacturing. 2019;**34**:463-474

[11] Panc NA, Bocaneț V, Vilău C, Chezan H, Popan IA, Conțiu G. New method of fixturing the low-rigidity parts by adhesive clamping. International Journal of Advanced Manufacturing Technology. 2023;**129**(5-6):2591-2608

[12] Yao S, Ozturk E, Curtis D, McLeay T. Work holding assessment of an UV adhesive and fixture design method. International Journal of Advanced Manufacturing Technology. 2020;**106**(1-2):741-752

[13] Biszczanik A, Talaśka K, Wilczyński D. Analysis of the adhesive spread and the thickness of the adhesive bonded joint depending on the compressive force applied to bonded materials with different surface structure. International Journal of Adhesion and Adhesives. 2022;**114**:103081

Chapter 6

Novel Wet-Field Mineral-Organic Adhesives Used to Repair and Regenerate Bone

George W. Kay, Rahul Jadia, Sourabh Boruah, Brittany McDonough and Michael C. Brown

Abstract

We describe the development and properties of a novel class of biomimetic and bioactive, mineral-organic, injectable, wet-field, high-strength bone adhesives providing new capabilities for surgeons. Tetranite has immediate load-bearing strength, durably bonding living bone and metal surfaces. Handling properties and biodegradation kinetics are widely tunable through multiple paths of formulation design. These calcium phosphate-based adhesives are bioactive, biodegradable, osteoconductive, and remarkably bone regenerative. The O-phospho-L-serine constituent, a common endogenous metabolite, has been shown to promote regenerative biological response, stimulating proliferation and phenotypic expression of osteoblasts. This unique bio-adhesive class is currently undergoing clinical trials in orthopedic and dental fields. The user-friendly properties are expected to improve clinical practice and patient outcomes by allowing new minimally invasive approaches, improved recovery times, reduced pain, and lowered cost of care.

Keywords: adhesive, biomaterial, regenerative, mineral-organic, self-setting, biocompatible, bone, biomimetic

1. Introduction

The search for novel adhesives frequently reaches into unusual and exotic places. This is the story which, in search of a bone regenerative wet-field adhesive, starts in the Pacific Ocean off California coast and is now continuing in human clinical trials involving the restoration of cranial continuity following craniotomies, dental implant stabilization, as well as preclinical studies of repair of long bone fractures, repairs of collapsed vertebrae, and bone void filling, among others. Even though the adhesive is a biomimetic derivative of the one evolved by the sea creature sandcastle worm *Phragmatopoma californica*, it is not related to the well-studied and described adhesive process using catechol oxidase and peptidyl-DOPA to attain adhesion [1, 2]. The biomaterial presented in the present chapter also does not rely on exotic polymers such as polyurethanes, [3, 4] cyanoacrylates [5], or foreign proteins [6] used by other bio-adhesives to achieve wet-field bone adhesion and does so without significant cytotoxicity, immunological risk, or other potential adverse effects. Additionally, no covalent bonds are either formed or broken during curing and degradation of the adhesive, thus eliminating the risk of exotic cytotoxic compounds leaching from the deposits of the biomaterial over its lifetime *in vivo,* eliminating both toxicity and carcinogenicity risks.

The novel adhesive class described in the chapter and branded Tetranite® is related to the calcium phosphate cements [7] but constitutes a class of its own because of its critical reliance on an organic component for function and surpasses their performance in several critical measures. The reactions involved in the adhesive performance are aqueous solution based. Both the behavior of the biomaterial to cure to a solid and its adhesive tack property are driven by chemical interactions of the ions present in the formulation. The fundamental constituents of Tetranite adhesive are a powder blend of multiphasic calcium phosphate salts, tetracalcium phosphate (TTCP) and tricalcium phosphate (α-TCP), and phosphorylated amino acid O-phospho-L-serine (OPLS) and water which initiates the reaction characterized by adhesive behavior [8]. All of the ionic species participating are endogenous to the mammalian body. In fact, OPLS is a common metabolite involved in both glycolysis and gluconeogenesis as well as multiple other anabolic and catabolic pathways leading to biosynthesis of such essential biomolecules as hemoglobin, glycine, serine, cysteine, and glutathione [9]. Furthermore, OPLS is a component of many proteins resulting from posttranslational modifications of native proteins by phosphorylation of the serine residue [10, 11]. In addition, OPLS has been shown to stimulate bone formation both *in vitro* and *in vivo* [12]. The handling properties as well as the behavior *in vivo*, e.g., setting time and rate of bone substitution, are controlled by small additions of pH and solubility modifiers as well as porogens to suit clinical requirements. Activation of the material generates a flowable and injectable adhesive which on the scale of minutes thickens into a workable putty and ultimately a rigid cement-like solid without clinically excessive exotherm.

The degradation of the biomaterial combines its passive thermodynamically driven physical dissolution in interstitial fluid with its active resorption mediated by multinucleated giant cells and other cellular actors. The removal of the adhesive is accompanied by neovascularization and osteoblast invasion of the deposit during a process which is substantially volume preserving. Although Tetranite adhesive is a robust bone regenerative biomaterial, it does not induce ectopic bone formation.

The characteristics of an ideal bone adhesive have been enumerated by Farrar [13] and others [14–18] and include multidimensional considerations for a diversity of clinical applications. A comprehensive listing of essential and desirable properties of a clinically relevant bone adhesive must include many of the attributes detailed below in **Table 1**.

This chapter traces the development and characteristics of a promising platform adhesive biomaterial technology, Tetranite, which satisfies all of the above listed requirements. Furthermore, the biomaterial offers sufficient flexibility in the central chemical processes driving its behavior to provide flexibility for a broad range of dental and orthopedic clinical applications. This flexibility can be exploited to gain significant differences in both mechanical and biological properties [19, 20, 24–27].

Both the bulk properties and the adhesive properties are a function of interactions among several multidentate species present within the formulation. The chemical interactions involved in the generation and curing of the adhesive readily fall into both Brønsted-Lowry and Lewis Acid-Base reaction models as well as electrostatic ionic interaction model. For example, Ca^{2+} ion is an electrophile capable of

Attribute	Practical reason and detail	Tetranite status
Wet field adhesion	Robust and tenacious bond strength to mineralized tissues and to potential implant surfaces without loss in requisite strength throughout bone substitution process	Yes, 3 MPa to high surface energy substrates including living bone, passive metals, glasses and ceramics with original formulation [19], greater in certain modified formulations
Biocompatibility	Non-immunogenicity, non-cytotoxicity, non-mutagenicity, non-interference of material or its degradation products with organ systems	Yes, passed applicable ISO 10993 biocompatibility tests, shown to be exceedingly biocompatible in multiple preclinical studies
Cohesion when applied clinically	Essential for adhesive strength of bulk material, Provides dimensional stability on placement,	Yes, resists washout in a wet field during application and maintains its integrity without migration, see Appropriate viscosity, below
Cohesion when cured	Essential for adhesive strength of bulk material, Provides lack of fragmentation and migration once solid	Yes, maintains its integrity without migration when subjected to loading after the material cures, see Mechanical Strength below
Mechanical strength	Prevents micromovements, persistent compressive and tensile strength, fatigue resistance, and fracture toughness	Yes, the immediate compressive strength upon curing is 3 MPa, which increasing with time ~10x the initial strength [20]
Appropriate viscosity	Low enough to allow injectability, high enough to control placement and confine biological response to desired location	Yes, 5000 centipoise rising over minutes until solidified, rise rate adjustable by product
Not infiltrated and replaced by fibrous tissue	Prevents formation of fibrous union and fibrotic invasion of grafted spaces.	Yes, replaced by osseous tissues under most conditions when in contact with bone, does not require guided tissue regeneration membranes
Biodegradable	Prevents long term foreign body issues and consequent complication risk	Yes, biodegradation when contacting bone adjustable weeks to years [20, 21]
Clinically relevant curing time	Clinical procedure and operation room time are limited by practical considerations	Yes, 1 min- 15 min adjustable by design parameters
Ability to serve as vehicle for drug administration	Antibiotics, pain control agents, growth factors, chemotherapeutics	Yes, incorporation, stability, and release demonstrated for antibiotics [22], bupivacaine, and others [23]
Sterilizability	Stability under conditions of either high heat or irradiation	Yes, sterilized by γ irradiation
Low exotherm process in clinical scale of use	Tissue damage can be induced by extended rise in local temperature related to polymerization, curing, etc.	Yes, in typical application volumes temperature does not exceed 41°C for clinically relevant time interval
Stability in storage	Practical shelf life	Yes, confirmed commercially viable shelf life
Radiopacity	Aids in monitoring of closed and minimally invasive procedures	Yes, the radiopacity of TN adhesives lies in the range between dense cortical bone and cancellous bone, but addition of $BaSO_4$ results in significantly more radiopaque material without significant modification of other properties
Reasonable cost of production	Ingredients, production cost, environmental impact	Yes, readily accessible raw materials and technology

Table 1.
Essential and desirable attributes of a clinical bone adhesive with specific reference to Tetranite adhesives.

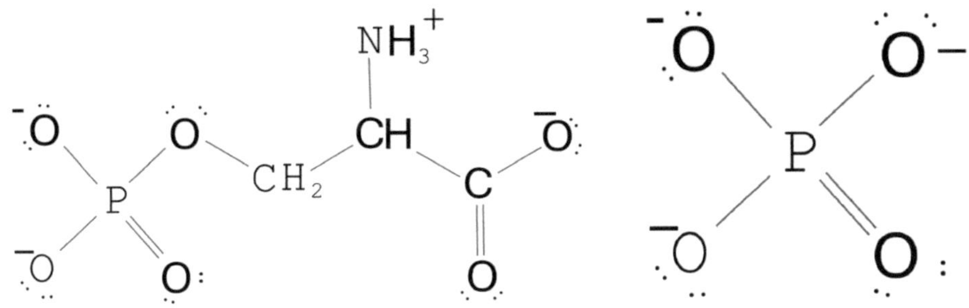

Figure 1.
Molecular structures of OPLS and orthophosphate ions.

coordinating six nucleophilic groups while also being a di-cation capable of interacting with multiple anions and dipoles. One of the crystalline reaction products, calcium phosphoserine monohydrate (CPS), is indeed stabilized by the coordination of the calcium ion by six ligands, all of which are nucleophiles, and some have net negative charge [28]. OPLS and orthophosphate, **Figure 1**, on the other hand, are polyanionic at physiologic conditions as well as nucleophilic by virtue of numerous lone electron pairs residing on the nitrogen- and oxygen-containing functional groups. OPLS possesses seven nucleophilic atoms capable of coordinating with electrophiles. Furthermore, they can form hydrogen bond-stabilized lattices.

Additionally, at physiologic conditions, ionized $OPLS^{2-}$ exists as a zwitterion, accepting a proton at its amino group nitrogen and thereby carrying a positive charge there, while the carboxyl group maintains a single negative charge and the phosphate group double negative charge [29]. Given the multitude and high complexity of the potential interactions, it is not surprising that multiple competing reactions take place at various successive phases following the activation of the system by provision of the water solvent.

2. Biological response: Biocompatibility

Given that the intended use of the adhesive discussed in this chapter is as a bone implantable material, Tetranite adhesive formulations have undergone the full ISO 10993-1:2018 [30] battery of applicable acute, sub-chronic and chronic biocompatibility measures and passed all, **Table 2**. Subcutaneous and intramuscular implantation of the material resulted in thin capsule formation without ectopic bone formation. Direct application to bone did not lead to loss of viable osteocytes in the immediately subjacent cortical bone.

Figure 2 demonstrates the benign response of living osteocytes in close proximity to two Tetranite formulation variant implantations, dense TN-SM at 10 days (a) and porous TN-ADBS at 6 months post-operatively (b). The adhesive has been mostly replaced by new bone tissue in the 6-month preparation.

3. Chemical processes and resulting physical changes occurring during activation and curing of tetranite adhesives

The activation of the adhesive consists of combining the powdered components and water, and, by doing so, providing a solvent for the process. The initial reaction,

Report name	Standard #	Pass/Fail
Chemical characterization	ISO 10993-18	N/A
In vitro cytotoxicity – MEM elution	ISO 10993-5	Non-Cytotoxic
Sensitization – GPMT	ISO 10993-10	Non-Sensitive
Irritation – intracutaneous	ISO 10993-10	Non-Irritant
Material-mediated pyrogenicity	ISO 10993-11	Non-Pyrogenic
Acute systemic toxicity	ISO 10993-11	Non-Toxic
Two-week muscle implantation	ISO 10993-6	Non-Reactive
Four-week muscle implantation	ISO 10993-6	Non-Reactive
Nine-week muscle implantation	ISO 10993-6	Non-Reactive
13-week muscle implantation study in rats with systemic toxicity	ISO 10993-6 ISO 10993-11	Non-Toxic
26-week muscle implantation study in rats with systemic toxicity	ISO 10993-6 ISO 10993-11	Non-Toxic
Bacterial Reverse Mutation (Ames) Assay	ISO 10993-3 ISO/TR 10993-33	Non-Mutagenic
In vitro mouse lymphoma assay	ISO 10993-3 ISO/TR 10993-33	Non-Mutagenic

Table 2.
List of ISO 10993-1:2020 biocompatibility tests [30] passed by TN-SM formulation.

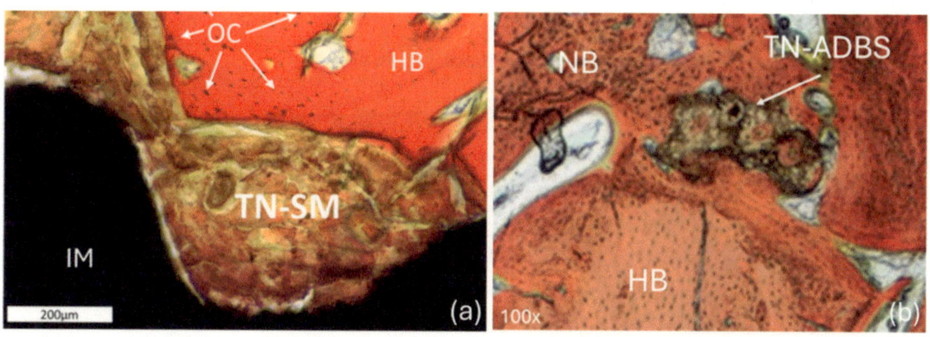

Figure 2.
Photomicrographs of Tetranite biomaterial implantation sites in contact with fresh osteotomy walls showing viability of osteocytes at short and long timepoints. (HB – host bone, pale red; NB – new bone, more intense red; OC – osteocytes; IM – implant stabilized by Tetranite adhesive. The adhesive appears mottled brownish. Canine mandibular residual ridge oversized implant osteotomy model at 10 days post-implantation (a) and segmental resection model at 6 months post-implantation (b). (Canine mandible, resin-mounted, nondemineralized ground section, Stevenel's blue, and Van Gieson picrofuchsin stain).

and the one releasing the successive cascade of events, is the ready dissolution of the highly water soluble and acidic OPLS, which is accompanied by a significant pH drop during the initial seconds of the process. The acidic medium then etches the surface of the alkaline but relatively insoluble calcium phosphate powders, TTCP and α-TCP, increasing the pH value of the reaction mixture to physiologic levels in the process. During this transformation from a water slurry to a slurry suspended in an increasingly concentrated solution of ionic species forming hydration shells as well as transient complexes, the change in the chemical state of the reaction mixture is matched by dramatic changes in the physical state which is marked by the rapidly increasing viscosity. This rise is illustrated by viscometry trace of the reaction mixture shown in **Figure 3**. Following activation and mixing, the adhesive was subjected to oscillatory shearing. The time at which the maximum tan (δ) occurs is at the cross-over of the loss modulus and storage modulus (G"/G'). This point is considered to signal the transition when the material has reached the most viscous response to deformation and has reached the end of its working time.

Furthermore, the increasing concentration of the solutes capable of forming coordination and ionic bonds in the face of a diminishing availability of free water tends to increase formation of chains and three-dimensional aggregates of these ions that are still solvated but relatively much larger than single molecules or atomic ions. Formation of these associations into microdomains is likely reversible early in the process, but as the free water is depleted, the growth of these enlarging domains promotes precipitation of progressively larger lattices and a transition into more viscous states and finally a putty and then a solid becomes irreversible. Based on our X-ray diffraction (XRD) analysis, the initial precipitate is primarily an amorphous highly hydrated lattice which possesses sufficient capacity for diffusion to allow for rearrangement of its structure depending on conditions. We believe that this behavior carries critical implications to the adhesive behavior of the material. The supporting evidence as well as a discussion of adhesion mechanism specifically is presented in the following sections. Physical state of the adhesive during the early curing can be described as a somewhat and increasingly viscous, cohesive, and tacky injectable suspension, approximating the consistency of thick honey, which can be manipulated with a syringe, spatula, or other instruments in a clinical setting.

3.1 Chemical phase evolution of tetranite adhesives

Rapid precipitation of salts, initially as a highly hydrated amorphous matrix of calcium phosphates and OPLS, leads to the formation of increasingly large domains of dynamic metastable three-dimensional lattices stabilized by ionic, coordination and hydrogen bonds. This process transforms the reaction mixture into a stiffening putty, whose adhesive properties diminish with time; however, this putty can be

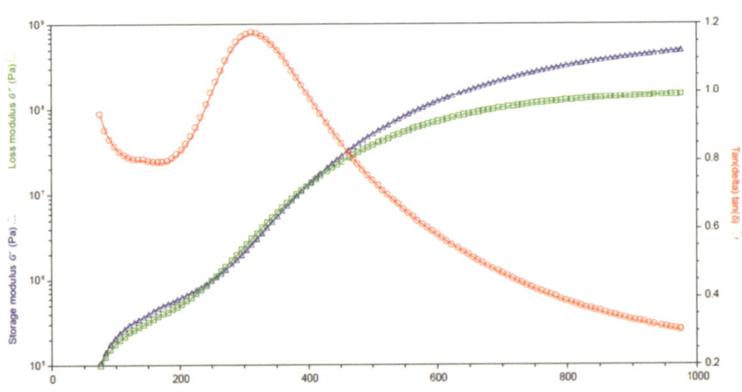

Figure 3.
Viscometrical characterization of the setting kinetics of TN-CFF adhesive. (Ares-G2 Rheometer, TA Instruments), 8 mm ø flat stainless-steel platens, 1 mm gap, 1,6 Hz, and 0,05% strain).

plastically molded and formed. After several more minutes, once it hardens into a brittle solid, the material may only be shaped subtractively, for example, by shaving. On a timescale of minutes to weeks, a spontaneous thermodynamically driven physical process is observed in immersed material *in vitro* without participation of biological agents. It is a rearrangement of the strained lattices within the precipitate to generate grain growth of micro crystallites of less strained domains. This is observed to first involve grain growth of calcium phosphoserine monohydrate (CPS) [28] and later of calcium-deficient hydroxyapatite [31] and other hydroxyapatite-like phases, within the amorphous domains and at the interfaces with fluid-filled spaces. The former appears to be kinetically favored and proceeds until the free OPLS is substantially consumed, while the latter is generated by hydrolysis of remaining TTCP starting material if sufficient free water is available. This chemical phase evolution of the adhesive continues for days and weeks both *in vitro* and *in vivo*. The reactions and rearrangements are governed by kinetic and other thermodynamic factors in addition to environmental factors and availability of reactants. Generally, less soluble phases are produced from the metastable ones as the rearrangement continues. **Figure 4** demonstrates the compositional change sequence as it is observed through XRD pattern changes in an immersed curing, and porous Tetranite formulation over time spanning from minutes to days following initiation of the reaction, that is, activation of the adhesive.

The predominant chemical reactions taking place in the system following activation follow Eqs. (1) and (2). The kinetically favored reaction consumes TTCP, OPLS, and water and results in the formation of calcium phosphoserine monohydrate, CPS, and phosphoric acid, Eq. (1). This reaction tends to rapidly consume free OPLS and ceases. One of the immediate consequences of this reaction progressing is a measurable drop in pH of the system.

$$Ca_4(PO_4)_2O + 3H_2O + 4C_3H_8NO_6P \rightarrow 4\left[Ca(C_3H_8NO_6P)(H_2O)\right] + 6H^+ + 2(PO_4)^{3-} \quad (1)$$

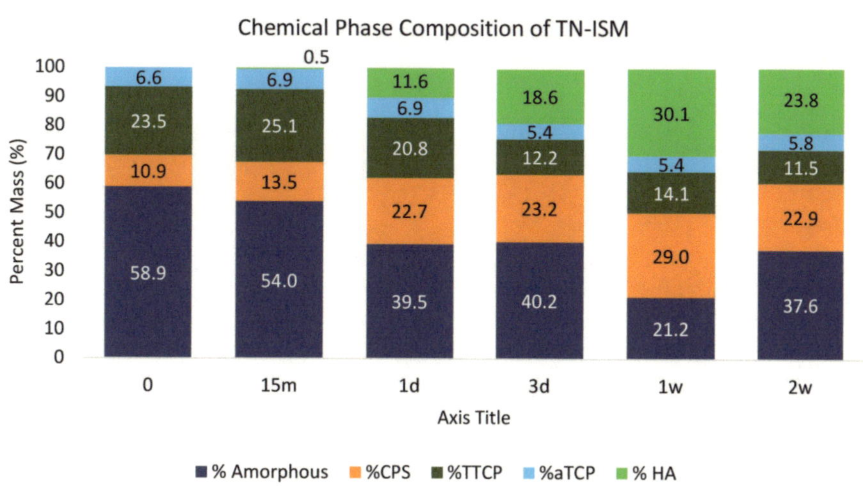

Figure 4.
Summary chart of continuing chemical phase evolution of the solid phases from the first minutes after activation to 2 weeks following it. Phase abundance determined by X-ray diffraction (XRD, ±10%, JADE, Materials Data, Inc.)

Eq. (2): A competing reaction is another process consuming TTCP and water. Albeit progressing slower, this reaction continues for days and generates hydroxyapatite and calcium hydroxide.

$$3\,Ca_4(PO_4)_2O + 3\,H_2O \rightarrow Ca_{10}(PO_4)_6(OH)_2 + 2\,Ca^{++} + 4(OH)^- \qquad (2)$$

As seen in Eq. (2), the reaction generates base, resulting in a rebound rise in the pH toward neutrality. Depending on the needs of the clinical application, the total titerable alkalinity/acidity as well as magnitude of the pH swing can be adjusted by the specific design of the formulation.

When *in vivo* and in contact with bone tissues, this physical process leads to strength increase and is augmented by ingrowth of mineralized bone tissue over the course of weeks to months. Such progressive formation of a biocomposite consisting of the remaining biomaterial increasingly penetrated and replaced by newly deposited mineralized bone tissue not only increases integration of the biomaterial by its fusion with the skeleton but also strengthens the deposit by incorporating increasing proportion of bone itself into the deposit. **Figure 5** illustrates the time course of substitution of the several original chemical phases of the adhesive with those of mineralized bone. The XRD data were collected from cores harvested at the indicated time points from mandibular residual ridge block defect sites (8mmLxmmH through-and-through) implanted with TN-ADBS and allowed to heal. The bone replacement process appears to be approximately half complete at the 3-month time point and substantially complete by the 5-month time point if one uses the native bone phase composition as the benchmark. It should be noted, however, that the crystallite composition of the mineralized bone, whether native or regenerated, is typically calcium-deficient and hydroxyl-deficient, carbonate-containing, and non-stoichiometric bone apatite and not stoichiometric hydroxyapatite identified by X-ray diffraction analysis because they sufficiently share crystalline lattice structure [32].

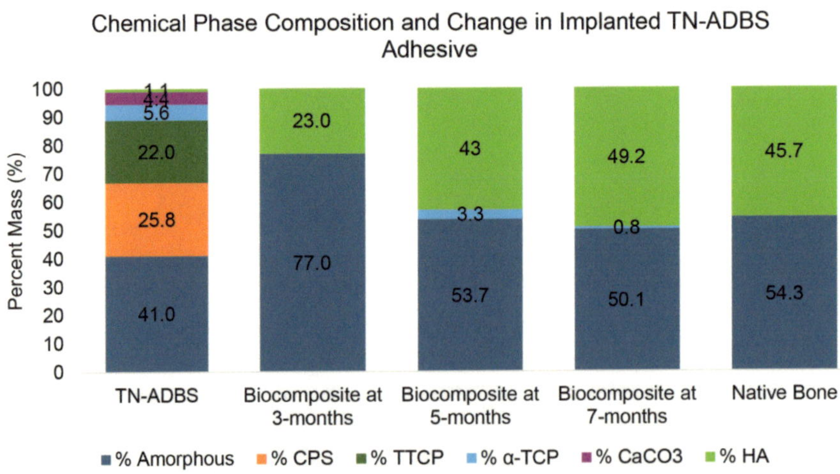

Figure 5.
Phase composition of highly porous formulation TN-ADBS of the adhesive and its evolution over the course of 7 months in vivo, *compared to phase composition of the implanted TN-ADBS adhesive and native bone harvested from the same anatomical site (Canine residual mandibular ridge model, TN-ADBS Lot# DEV23012; XRD analysis, JADE, Materials Data, Inc.).*

3.2 Early wet-field adhesion behavior of tetranite

During the early phases of curing that follow the activation of the adhesive, the reaction mixture exhibits significant tack to high energy surfaces, including living bone, passive metals, glasses, ionic salt solids, structural proteins, and others. This behavior, a composite of chemical adhesion, cohesion, and viscosity has been characterized *in vitro* by tack testing. The time dependence of this behavior is illustrated by **Figure 6**.

The combination of attributes of wet tack, injectability, and high degree of biocompatibility characterizing Tetranite adhesives during early curing may be exploited clinically for fixation of bone fragments or stabilization of implants in bone as they are unique among biomaterials, particularly when combined with persistent load-bearing strength, bioresorbability, and highly favorable bone regenerative properties that these adhesives exhibit [33–35]. The duration of the flowability and adhesive behavior window depends on the specific formulation version and may be adjusted in a range from seconds to several minutes depending on clinical requirements. Likewise, the rate of *in vivo* resorption of the material can be engineered from approximately 10 days to over a year. Even though the ability to form new adhesive bonds ends as the physical state changes to that of a putty, the bonds formed during the earlier phase not only persist but also become more robust.

3.3 Chemical basis for adhesion of tetranite reaction mixture

At the chemical level, all evidence of the mechanism of adhesion and cohesion related to this reaction points to three types of bonds. These are ionic bonds, coordination bonds, and hydrogen bonds. The polydentate zwitterion OPLS, with its amino, carboxy, and phosphonooxy functional groups, each providing a charge for ionic bonding as well as nucleophilic or hydrogen bonding sites, is a primary bridging anionic molecule [29]. Calcium cation (Ca^{2+}), with its double positive charge and the ability to simultaneously form coordination bonds with up to six nucleophilic

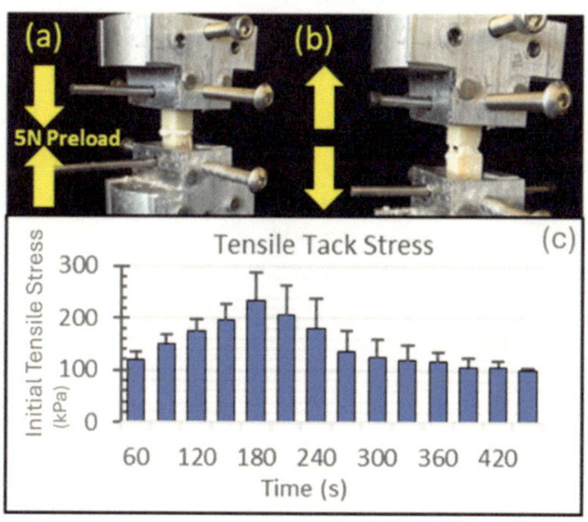

Figure 6.
Tack behavior exhibited by Tetranite TN-SM adhesive. Apparatus used to measure tack during cyclical compression (a) and distraction (b) between cortical bone block surfaces. Time course of tensile stress results as a function of time (c) (Instron Model 5969 testing machine, crosshead speed in tension 0,5 mm/second, cycle length of 30 sec, 2 mm displacement per cycle).

ligands, is the primary cationic bridging species, both as an electrophile and a highly charged ion. The multiple orthophosphate ionic species as well as water molecule are additional coordinating and hydrogen bonding participants in forming the metastable precipitate lattices which makes up the bulk of the amorphous phase of the freshly cured adhesive. Corresponding ionic, nucleophilic, and electrophilic groups contributed by host mineralized and soft tissues provide substrate for adhesive bond formation as the water of hydration is stripped from these surfaces [14]. It is at the stage immediately following activation when the internal bonds have not yet irreversibly formed that the adhesive behavior is expected to be near its maximum.

There is evidence that the uncommon wet-field adhesion attribute of Tetranite formulations is related to the activity of water within its curing reaction. Generally, moisture tends to severely diminish adhesion effect in many otherwise adhesive systems. In wet environments, surfaces form hydration layers which typically include hydrogen, coordination, and other bonding between the water of hydration and the functional groups on the substrate surface. These interactions decrease the surface energy and act as weak boundary layers, thereby significantly reducing the interactions necessary for adhesion [14]. We postulate that the water deficient nature of the reaction mixture is responsible for its wet-field adhesive behavior by competitively, and perhaps even quantitatively, stripping water from the hydration layer and making the functional groups of the substrate surface, whether electrophilic, nucleophilic, or ionic, available for bonding with the several multidentate components of the activated Tetranite adhesive reaction mixture. As one of the supporting lines of evidence, we present the comparison of XRD patterns of dry cured-dry aged and dry cured-wet aged, crushed Tetranite (**Figure 7**). In preparation, the samples were crushed after reaching cure at 5 min post-activation. Note the chemical composition difference between the samples resulting from difference in access to water and its imbibition. The difference in composition is taken as evidence of the curing reaction not proceeding to completion, as it would *in vivo*, under conditions depriving it of water. Both Eqs. (1) and (2) indicate water as a reactant.

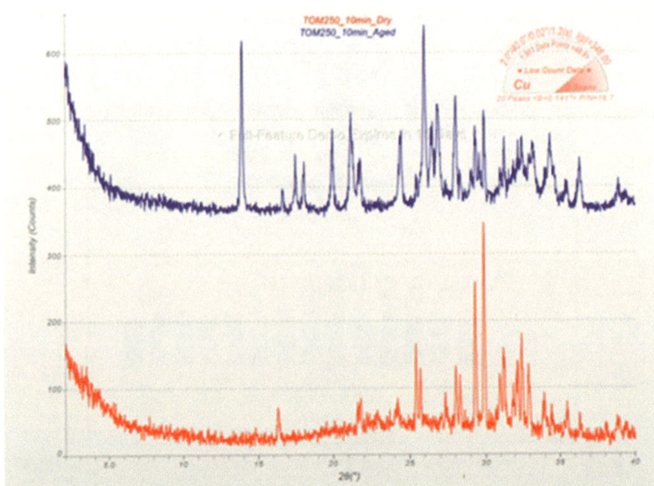

Figure 7.
Comparison of XRD patterns of TN-SM adhesive cured under identical conditions but aged for 10 minutes either in presence of excess water (top) or dry (bottom).

What is readily evident is the marked difference in the chemical phase composition of the two samples, one aged in a humid but not wet environment and the other provided with unlimited water. The conclusion drawn is that the continuing evolution of phases in the adhesive is dependent on ready availability of water that is consumed in processes involved in such evolution. In the context of the curing adhesive presented with a hydrated surface, the hypothesized model for the adhesive process is that the hydration layer water is diffused away from the substrate surface into the interior of the adhesive mass where it is competitively reacted or bound, thus allowing the formation of more stable bonds at the interface of the adhesive and the substrate. In the meantime, within the bulk of the active reaction mixture, the multidentate components continue to form bonds that lead to the formation of a lattice of the amorphous matrix and the microcrystalline domains within it.

4. Bond strength

In contrast to calcium phosphate-based cements, Tetranite adhesive formulations exhibit robust chemical interactions, that is, bonds, with substrates. It has been observed that the strength of the adhesive bond formed with high energy surfaces, such as polished faces of titanium or cortical bone blocks, increases over time *in vitro* on the scale of hours to days. **Figure 8** illustrates this phenomenon for bonds formed by TN-SM formulation between pairs of wet cortical bone and wet titanium blocks pressed together with 0,1 N load until testing.

This behavior is observed both *in vitro* and *in vivo*, suggesting that the process is intrinsic to the material and likely associated with the rearrangements occurring within it rather than through action of the host tissues. It might be postulated that these solid-state rearrangements tend to move in the direction of maximizing the total stabilization energy of the system by minimizing strained bonds and maximizing the number of stable bonds, which would lead to increased mechanical strength of the system. Additionally, the bond is readily formed on contact

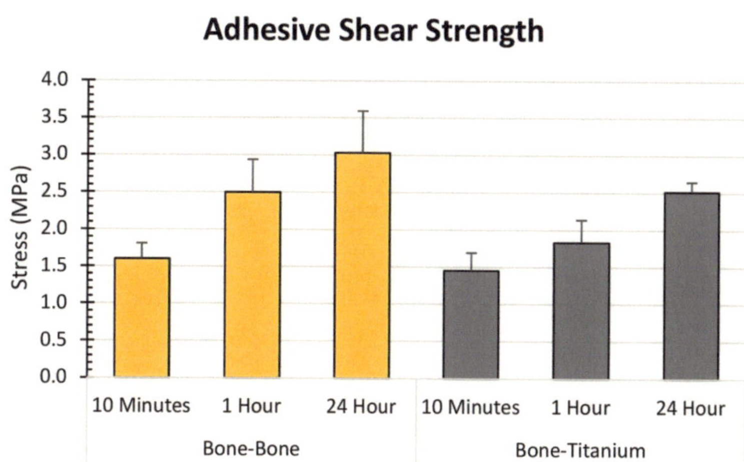

Figure 8.
Time dependence of adhesive shear strength of Tetranite adhesive bond to bone and titanium illustrating the wet field bond formation and its persistence.

with wet surfaces, even surfaces submerged in aqueous media such as blood or other tissue fluids, which particularly extends the utility of this class of adhesives to clinical procedures. Examples of this phenomenon, as observed *in vivo,* are presented below.

4.1 Biological and mechanical evidence for *in vivo* bond formation and strengthening over time

Two separate and distinct processes of chemical and physical bond formation are evident when Tetranite adhesives are implanted *in vivo*. They both demonstrate the bioactive character of these biomaterials [36]. The first process is active chemical bond formation by the adhesive itself to the tissues during its minutes-long flowable reaction mixture phase and is primarily a function of the chemical reactions ongoing within the adhesive itself and at its surface. The second, that of bone deposition directly onto the Tetranite surface, is only evident after days, weeks, or months following its cure and results from the biological response of osseous tissues to the implanted bioactive material. Even though all of the tissue histology images included within this chapter present direct evidence for one or both of these processes, we point to **Figure 2** along with the biomechanical data presented herein as the clearest evidence of the first process, and the high magnification photomicrographs presented in **Figure 9** demonstrating the second. Both of these processes contribute to the observed macro-scale increase in bond strength observed *in vivo*. It may be reasonably assumed that invasion of the material deposit by new bone, a strong and tough structure, occurring during the second process and the increasing surface area and complexity of the shape of the interface with the material simultaneously undergoing biodegradation are the primary drivers of the increase in mechanical strength of the construct *in vivo*. The photomicrographs in **Figure 9** present a view of grains constituting the adhesive (TN) and of invading bone tissues which include marrow and vascular spaces. Photomicrograph (a) illustrates the relationship between the remaining multiphasic grains of the biomaterial and the mineralized bone (mb) and marrow and vascular spaces occupying the intergranular regions from which the adhesive has been removed by biodegradation of less stable regions. The higher magnification field (b) displays an edge of a marrow space with mineralized bone (mb), osteoid (os),

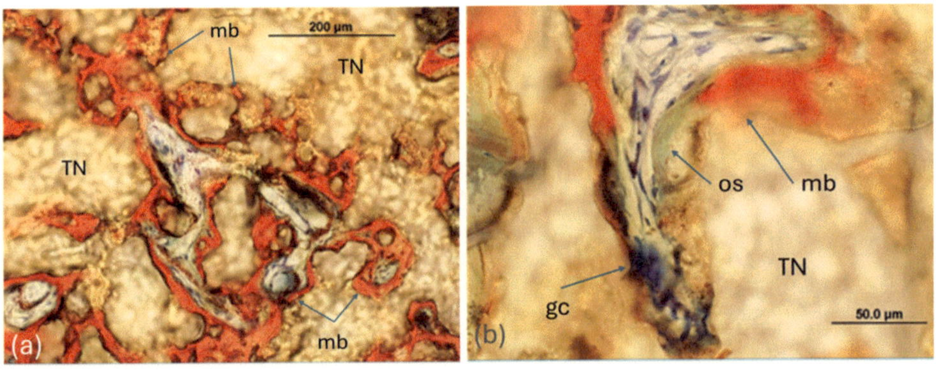

Figure 9.
Photomicrographs of dense Tetranite TN-SM formulation deposits undergoing degradation and direct bone deposition (canine mandible, 4 months post-implantation, resin-mounted nondemineralized ground section, 4 months post-implantation, Stevenel's blue, and Van Gieson picrofuchsin stain).

and a multinucleated giant cell (gc) evident on biomaterial surface. Note direct and distributed deposition of both the osteoid and mineralized bone onto the Tetranite surface.

The reaction products comprising the cured adhesive deposit are relatively stable *in vivo* on the clinically relevant time scale of weeks to a year. This allows for significant load-bearing strength of both the adhesive bond and the bulk material to persist while the material is first osseointegrated and then gradually replaced by bone. *In vivo*, because of the presence of multiple biological factors potentially affecting the material, the behavior of the adhesive is much more complex. An *in vivo* example of significant initial stability of a Tetranite adhesive-bonded construct which increases over time can be illustrated by the comparison the mechanical properties exhibited in a rabbit experimental spine fusion model (**Figure 10**) [37]. The output measures presented are the tensile rupture force (black bars) and the quotient of force required to rupture the experimental site divided by the force required to rupture adjacent intervertebral ligaments (blue line) in explanted and mounted specimens. The chart also presents for comparison data previously published for autogenous graft fusion (red line) by the author of the model [37]. Once again, the significant initial strength of the adhesively bonded construct, one equivalent to 6 and 10 weeks following fusion with autogenous bone, is followed by a multifold increase in strength over weeks following the adhesive treatment. In this case, the rupture strength magnitude is nine-fold greater than that of the adjacent intervertebral ligaments, which is more than two-fold improvement over autogenous bone fusion. Also of note is the remodeling of the adhesive material mass and the adjacent vertebral bodies which occurred during the nine-month course of the study and is evident in the 3-D reconstructions.

Among potential applications currently undergoing clinical trial is the use of Tetranite adhesive, TN-CFF, to replace titanium plates and screws to fixate cranial flaps following neurosurgical procedures. Even though plates and screws have remained the standard of care for decades because they provide immediate stability and alignment of the cranial flap, they are not designed to integrate with the

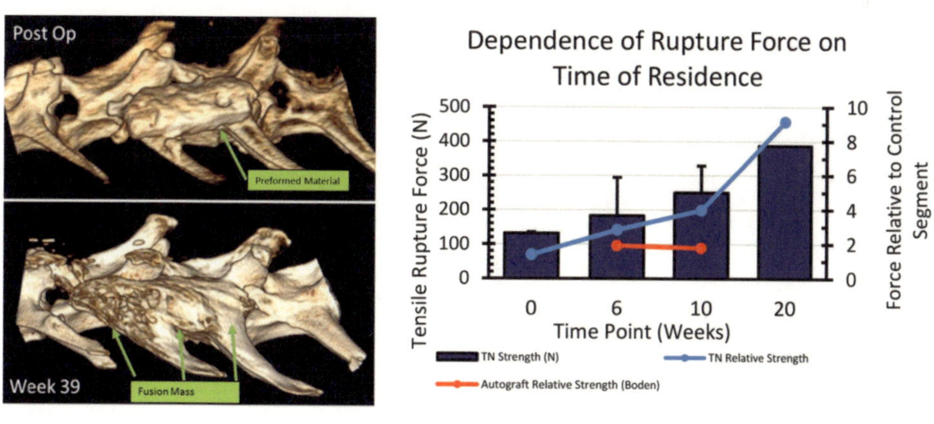

Figure 10.
3-D reconstructions of in vivo *longitudinal CT data from an exemplary site (left) showing spinal fusion by posterolateral approach with dense Tetranite adhesive stabilization of preformed Tetranite precast blocks. The upper image presents immediate post-operative appearance, while the lower was captured at the 39-week time point. The chart on the right shows the time dependence of the tensile force required to rupture the construct (left scale) and ratio of such tensile force relative that required to rupture of adjacent intact intervertebral joint (right scale). Comparison of Tetranite adhesive bond at* t = 0, *6, 10, and 20 weeks (blue line) with published autogenous bone grafted control (red line) [37].*

surrounding bone or promote long-term bone regeneration. This limitation has contributed to significant complication rates, which include infection, cerebrospinal fluid leakage, hardware migration, and poor cosmetic outcomes for patients [35]. Tetranite adhesives have shown encouraging results in studies in an *in vivo* ovine model reaching 2-year time points [21]. A more recent study evaluated bone healing through 12 weeks postoperatively. The study included 12 skeletally mature crossbreed sheep, each undergoing bilateral parietal craniotomies (**Figure 11a**). One side was repaired using TN-CFF, while the other was stabilized with titanium plates and screws (**Figure 11b**). The study conditions included a durotomy-and-repair group to reflect real-world surgical scenarios. The outcomes assessed included cranial flap volume, bone regeneration across the kerf line, and histological evaluations of local and systemic effects. Post-operative imaging and analyses were performed immediately and at 12 weeks to capture both short- and long-term results.

Cranial flap volume, an indicator of structural preservation, was significantly greater in TN-CFF-treated flaps compared to those stabilized with titanium plates and screws. At 12 weeks, the average flap volume in the TN-CFF group was 3.27 ± 0.82 cm^3 at post-op and 3.43 ± 0.67 cm^3 at termination, compared to 2.19 ± 0.53 cm^3 at post-op and 2.57 ± 0.89 cm^3 at termination in the titanium group, highlighting TN-CFF's ability to maintain flap integrity over time. Similarly, TN-CFF-treated sites exhibited significantly greater bone regeneration across the kerf line, with kerf bridging exceeding 86% at 12 weeks, compared to 61% in the titanium group (**Figure 11**). These findings were supported by histological analyses, which showed robust deposition of mineralized bone within the TN-CFF matrix and seamless integration with the surrounding

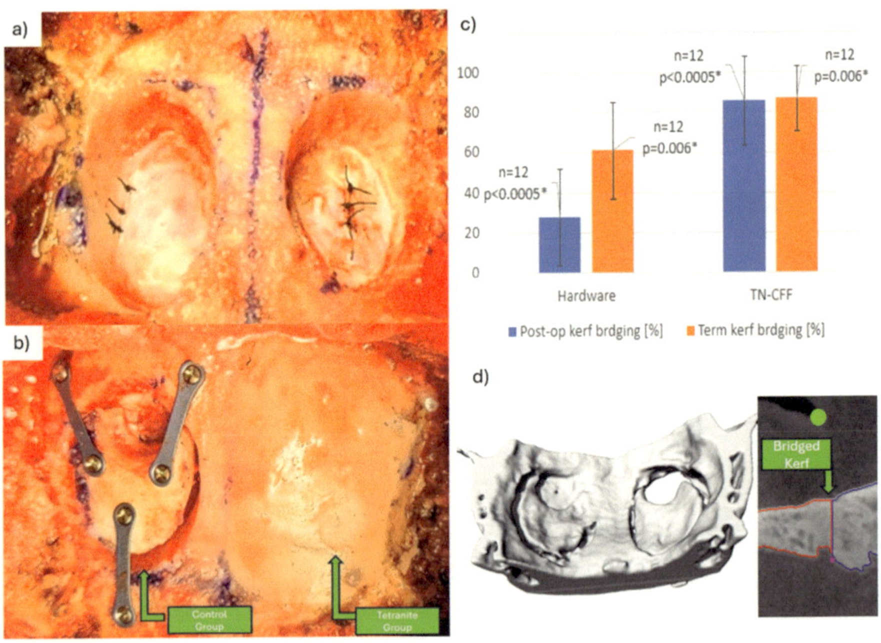

Figure 11.
Left: Intra-operative photographs showing bilateral craniotomy with repaired experimental dural tears (a) and fixated bilateral craniotomies with TPS (left) and TN-CFF (right) (b). Right: Chart showing percent kerf bone bridging at 12 weeks, measured by CT (c) and 3D reconstruction of CT data immediately post-op showing TN-CFF filling the kerf line and perforator hole (left) and an empty kerf line and perforator hole in the control (right), while 2D reconstruction of CT imaging at 12-weeks shows kerf bridging in the TN-CFF group.

cranial bone. Histological evaluations revealed no significant adverse effects attributable to the adhesive. Localized cortical changes were minimal and procedural in nature, unrelated to the treatment. Importantly, no evidence of systemic toxicity was observed, affirming the biocompatibility of TN-CFF. The adhesive's controlled resorption and gradual substitution by new bone ensure that its presence is transient, reducing the risk of long-term complications and eliminating the need for additional hardware removal surgeries.

By eliminating the need for hardware, TN-CFF could reduce the risk of CSF leakage, hardware migration, infection, and cosmetic dissatisfaction, all common reasons for revision procedures. Furthermore, its osteoconductive capabilities restore the native cranial anatomy, ensuring long-term functional and structural integrity. For patients, these benefits can translate to faster recovery, fewer hospitalizations, and improved quality of life, which is being evaluated in pending clinical studies in the United States and United Kingdom.

Another potential clinical application using Tetranite adhesive, TN-ISM, involves the stabilization of titanium hardware, dental implants, in a patient. The adhesive bonds providing implant fixation to the bone prove to be stable. Not only do they persist as long as the material remains before being totally replaced by osseointegrating bone tissues but also they maintain and enhance their strength. In **Figure 12** are presented photomicrographs of endosseous dental implants affixed to bone in the course of a preclinical study of implant stabilization by a dense Tetranite TN-SM adhesive formulation with a course of resorption of 1 year in the canine model [20]. **Figure 13** presents biomechanical data from the same study. The values presented in the chart are the level of reverse torque required to rupture the bond initially between the implant and the adhesive, but later, the biocomposite consisting of the adhesive and the bone penetrating it and the implant.

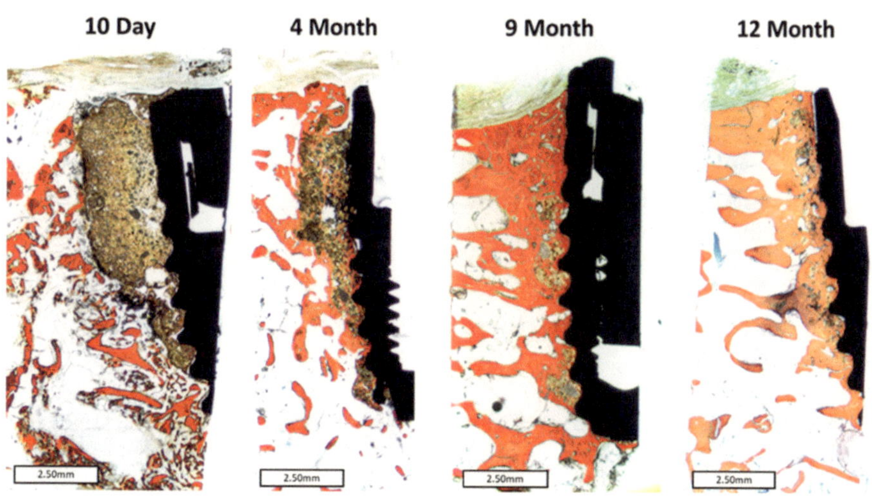

Figure 12.
Photomicrographs of Tetranite TN-SM stabilized dental implant sites at 10-day, 4-, 9-, and 12-month time points following implantation. Note the osteotomy design minimizing bone engagement by limiting it to the apical 2 mm of the implant. Also, note the dendrite-like penetration of the material by bone (red) to the implant surface, resulting in osseointegration, and progressive replacement of the biomaterial by new bone. (Straumann 3,3mmøx 8mmn bone level tapered implant, canine mandibular residual ridge oversized osteotomy model, sagittal plane, original magnification 25x).

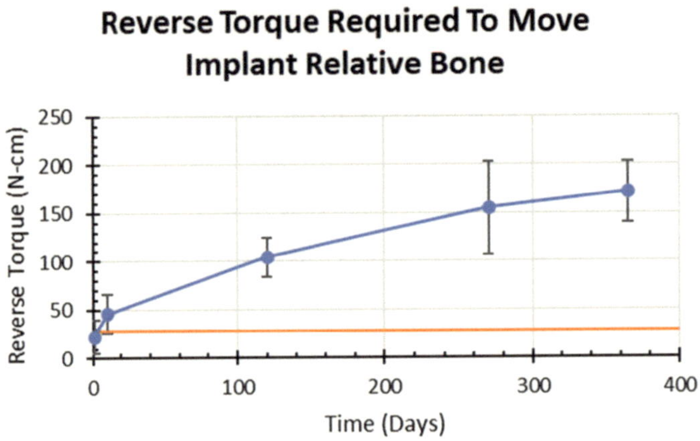

Figure 13.
Time dependence of removal torque of submerged dental implants adhesively stabilized by Tetranite in sites lacking primary stability in vivo. *Reverse torque of 30 N-cm (red line) is widely considered by clinicians to be an indication of sufficient stability for implant restoration (Canine mandibular residual ridge oversized osteotomy model, Ø3,3x8mm bone level tapered implant, Straumann, Basel, Switzerland) Statistics: 24 hours,* n = 6; *10 days-12 months,* n = 5.

Note that the initial stabilization, in this study measured by resistance to reverse torque, reaches clinically acceptable levels of >30Ncm when implants are initially stabilized with the bone adhesive during placement. However, the resistance of stabilized implants to removal torque rises further without stability loss experienced with traditionally placed dental implants during the time window when primary stability is diminished but secondary stability has not yet significantly developed [38]. It is clear from the photomicrographs that osteointegration and bone penetration of the adhesive proceed to full osseointegration of the implant itself and likely contribute to the increase in the measured removal torque early in the process. Note that there appears to be an asymptotic limit being approached. We believe that this limit corresponds to the time when the biomaterial is totally replaced by mature functional bone. Several human clinical trials using a more rapidly substituted, less dense Tetranite adhesive formulation TN-ISM are currently pending internationally. Results from pilot studies were encouraging [34, 39], and preliminary results from the pending larger pivotal studies in humans are favorable, but conclusive data are not available as of the date of this writing as patients are still being enrolled.

5. Modification of the adhesive formulation to meet specialized clinical requirements

It is clear that each anticipated clinical application may require a distinct range of attributes for optimal performance, necessitating the development of distinct formulations of the biomaterial custom designed for its purpose. The list of potential clinical applications of Tetranite adhesive are numerous and diverse, including dental implant stabilization, pedicle screw stabilization, cranial flap fixation, fracture fixation, vertebroplasty, spine fusion, arthrodesis, bone void filling, bone volume augmentation, and many other specialized uses. Some of the listed applications may use Tetranite as an intraoperative aid, while others may constitute definitive treatment, and others yet may be used in minimally invasive closed approach. As a direct consequence, handling variables (e.g., amount needed, setting time, mixing method, viscosity, etc.), physical properties (e.g., adhesive strength, compressive strength, porosity, presence or lack of expansion during curing, etc.), and biological properties (e.g., rate of resorption and bone substitution, bone regenerative potential of the material, inclusion of therapeutics such as antibiotics, anesthetics, growth factors, etc.) all need to be optimized for specific applications. For example, bulky bone void filling applications which are not load-bearing may require very different properties from a long bone fracture fixation treated by injection during a minimally invasive technique. Several of these potential modifications, and their impacts, are considered below.

Additives that affect either the properties of the primary product of the reaction, i.e., the precipitate itself, or the course of the reaction, for example, the setting kinetics, include ions that interact with the primary components or substitute for them. Among the cations are alternative multivalent metal ions, for example, alkaline earth elements such as magnesium [40] and strontium [41, 42] which are known to affect biological response of the host tissues, often in complex beneficial ways, when used as minor biomaterial components. Among potential candidate anions are multiple inorganic and organic acid conjugate bases, such as silicate [43, 44] and citrate [45]. Citrate naturally constitutes approximately 1–2% of bone mass, and of that 65–80% is tightly bound to mineral bone apatite, while the remainder is tightly bound to the collagen component [46] present within bone tissues.

A range of modification strategies have been employed to generate different versions of the Tetranite adhesive formulations for different clinical applications. These vary from inclusion of different components, such as a porogen, e.g., calcium carbonate, or radio opacifier, for example, barium sulfate, to choosing the packaging and mixing method that is appropriate for the specific application. For example, the alkaline earth element, barium, which is toxic to tissues when in solution, has been added as an insoluble and non-toxic sulfate salt, $BaSO_4$, [47] at 2.5% w/w to serve as a biologically passive radio-opacifier in Tetranite adhesive formulations which might be used through injection in closed procedures (**Figure 14**).

The methods of use also dictate the packaging and handling methods. For example, the dental implant stabilization version of the adhesive which is currently in clinical trials, TN-ISM, is a porous, slightly expanding, and rapidly resorbed adhesive which is packaged in need-matched doses of 0.5 ml and is mixed in and dispensed

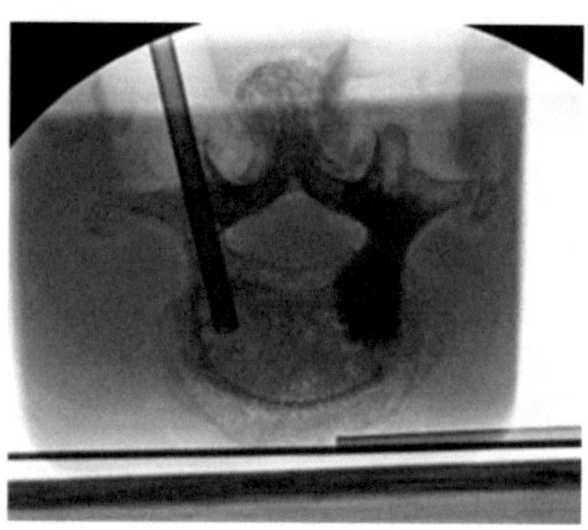

Figure 14.
Transpedicular minimally invasive injection technique using barium sulfate radiopacified TN-MFF adhesive under radiographic monitoring (human cadaver). The image illustrates the placement of the cannula on contralateral side (left) from completed injection (right).

from a sterile modified dental triturator capsule, commonly used in dentistry in a clinical setting. By contrast, the non-expanding and slower-resorbed cranial flap fixation adhesive TN-CFF, also in clinical trials, is packaged in 10 ml doses, mixed in a provided bowl, and applied with a provided delivery syringe. Examples of specific modifications of the formulation developed to satisfy specific clinical performance requirements are presented below.

5.1 Accelerated bone substitution

In certain clinical situations, a premium is placed on persistent high strength, while in others, the strength of the material is not critical, but more rapid bone substitution leads to improved clinical outcomes. The modification strategy used in the example provided is to make the bulk not only more soluble but also more porous, ~20-35%, thus enhancing the speed of bone replacement. Inclusion of sodium ion in the formulation increases the solubility by not only providing an ion generating less-stable solids but also one that strains the precipitated lattice. Additionally, a porogen, calcium carbonate ($CaCO_3$), can be added to release carbon dioxide (CO_2) bubbles during the early curing phases while the adhesive is still a cohesive liquid. Later, once the adhesive has solidified and hydrated, the gas diffuses out into the interstitial tissue fluid and is replaced by it. Another benefit of addition of carbonate is the incorporation of the ion into the precipitate forming during curing of the adhesive. This results in the precipitation of apatite solid containing substitutions of carbonate in either one of the phosphate positions or the hydroxide positions within the apatite lattice structure [48]. Carbonate inclusion yields a more soluble precipitate [49], leading to accelerated turnover to bone. Even though hydroxyapatite crystalline lattice is stable in the physiological environment of the body, substitution by sufficient foreign ions destabilizes the HA lattice.

Figure 15 presents the difference in phase composition of an unaltered dense legacy adhesive and a modified version, formulated to not only be porous but also to

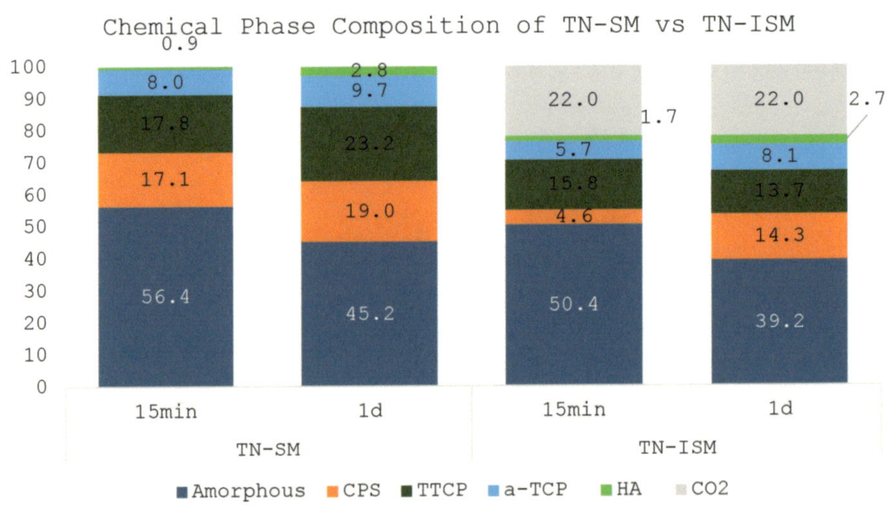

Figure 15.
Volumetric phase composition within a dense, TN-SM, and a porous, TN-ISM, formulations at 15 min and 1 day post-activation. Note the 22% gas (CO_2) content of the TN-ISM cured material. (n = 3 *XRD analysis, ±10%, JADE, Materials Data, Inc.)*

be more soluble in order to undergo accelerated bone substitution in clinical use. The goals of the design change to increasing surface to volume ratio for interaction with the tissue fluids and to generate less stable solid by introduction of straining ions into the lattice. The original dense formulation, TN-SM, contains only the primary active components, i.e., OPLS, calcium phosphate, and water, while the modified, TN-ISM, additionally contains NaOH and $CaCO_3$, with the carbonate ion at approximately 1/20th of the molar abundance of orthophosphate ion. It is worth nothing that while TN-SM gradually evolves to a more highly crystalline state accompanied by diminution of the amorphous content, that is, gains stability, TN-ISM remains substantially amorphous and less stabilized by the bonds of the crystal lattice because of the Na^+ and CO_3^{2-} substitution. It should be noted that XRD methods might not be sensitive to the lattice distortion at the doping level used.

The rabbit distal femur critical size defect model [50] has been extensively used for the assessment of biological response of osseous tissues to candidate formulations of Tetranite adhesives. **Figure 16** demonstrates the results of implantation of dense original formulation, TN-SM, and a modified variant, TN-ISM, differing by the inclusion of 2% (w/w) $CaCO_3$ porogen among the solid components as well as inclusion of NaOH in the formulation. The seemingly small formulation change produces a dramatic difference in the physical state, as seen in **Figure 10** and in the biological response, as seen in **Figure 16**. At 3 weeks following implantation the substitution with bone in of the unmodified adhesive is proceeding only at the surface of the solid deposit and without detectable adverse reaction or much penetration into the interior of the deposit (a). It should be stated that this original formulation was used in the examples of *in vivo* use in implant stabilization (as thin layer, **Figure 12**) and spinal fusion (as bulky mass, **Figure 10**) discussed above. In contrast, the porous deposit at the same time point (b), not only has been deeply penetrated by bone tissue ingrowth, but it is also substantially replaced by newly generated bone tissues. Of further note, the TN-ISM formulation is currently tested in a multinational clinical trial of its safety and efficacy for stabilization of dental implants in sites failing to provide primary stability.

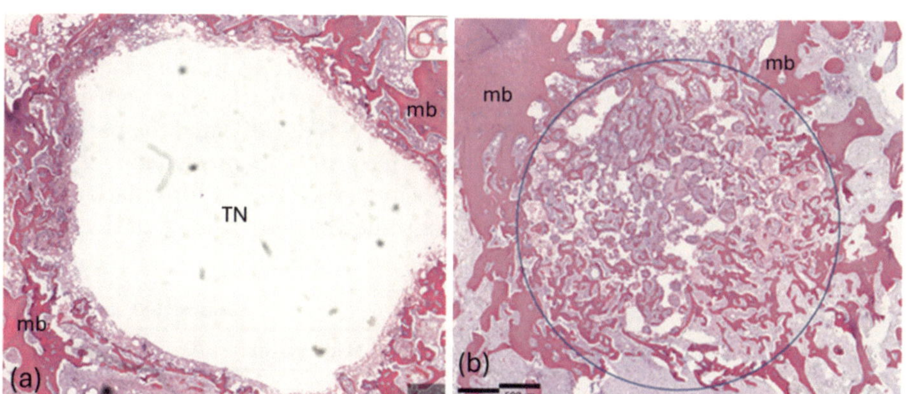

Figure 16.
Photomicrographs illustrating relative rates of penetration and replacement by bone of two Tetranite adhesive formulations, TN-SM (a) and TN-ISM (b) 3 weeks following implantation. White fields represent the space remaining after the biomaterial was dissolved away during demineralization of the sample in the course of slide preparation. A circle demarcates the original interface between host bone and 5mmø x 10 mm cylindrical deposit of the adhesive. (Rabbit distal femur critical size defect model, demineralized, paraffin mount, H&E stain).

5.2 Further porosity considerations

Porosity of biomaterials and its influence on the biological response of osteoblast cell line has been extensively studied [51–55], and it has been found that the level and characteristics of porosity can be central to the rate of interaction of bioactive materials. At the most superficial level, while a dense solid interacts with the host only at the envelope surface exposed to the tissues, a solid penetrated by open and interconnected porosity interacts through a greater surface area which includes deeper regions of the interior. This presents greater surface area for tissue interaction while reducing the mass of material to be degraded. That interaction is not only dependent on the percentage of porosity within the biomaterial but also on its shape, size, and surface chemistry. **Figure 17** demonstrates the current level of control within the Tetranite family of biomaterials for both the size and shape design variables. Presented are images representing two dimensional reconstructions of micro-computed tomography (μ-CT) data of three Tetranite formulation variants incorporating different porosity levels, types, and dimensions. The material on the left, TN-SM, is dense, with porosity content of 1.4%. The formulation variant in the center image is TN-ISM, discussed above, containing 23.6% closed porosity. The variant on the right, TN-ADBS, in addition to calcium carbonate to the adhesive binder contains a substantial admixture of precured substantially porous Tetranite granules. The volume of the binder is insufficient to fill all the intergranular volume, resulting in a high level of interconnected open porosity, 32-60%, which can be engineered to desired pore size. **Table 3** provides additional quantitative details regarding porosities obtained in the formulation variants presented in **Figure 17**.

The two sets of data presented are as follows: 1) all porosity resolved at 18 μm resolution and 2) all porosity algorithmically filtered so that only pores greater than 100 μm in diameter are counted.

In a canine study, dental implants were immediately placed into mandibular tooth extraction sites following aggressive curettage of the cribriform plate and oversized osteotomy only engaging the implant at the apical 2 mm. During implant placement, insertion torque levels of less than 15 Ncm confirmed lack of primary stability. These implants were then stabilized by application of the indicated bone adhesive formulation to provide stabilization. Loosening of the implants stabilized by the biomaterial by application of reverse torque-to-failure was used as a proxy measure for functional adhesive strength of the modified formulations. Two formulations presented in **Figure 17** were compared: the dense formulation, TN-SM, (left) and the porous formulation, TN-ISM, (center). The maximum torque results are presented in **Table 4**. Given the greater than tenfold amount of porosity in the TN-ISM formulation relative the dense TN-SM adhesive formulation, it was somewhat surprising that the removal

Formulation	All Porosity			Porosity >100 μm		
	Median φ (μm)	Pore count	Porosity (%)	Median φ (μm)	Pore count	Porosity (%)
TN-SM	22,5	13,458	2,67	124	157	1,44
TN-ISM	33,6	33,680	29,93	132	2421	23,70
TN-ADBS	30,7	12,276	34,80	126	385	32,88

Table 3.
Comparison of descriptive statistics of porosity as measured in the three Tetranite formulation variants shown in **Figure 17**.

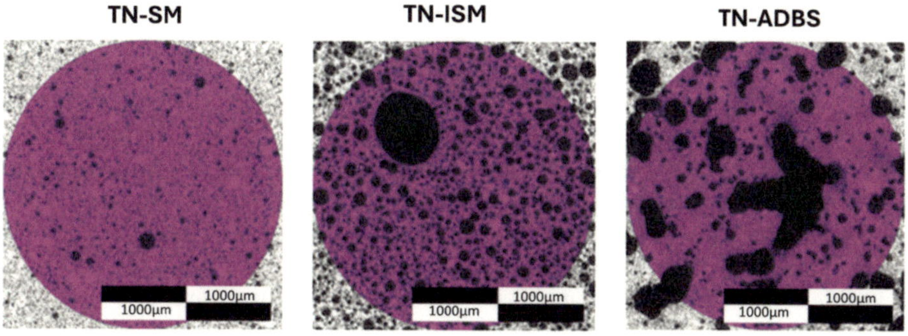

Figure 17.
Two-dimensional reconstructions of μ-CT data of three formulations exhibiting different porosity levels and types: dense formulation TN-SM (left), closed-porosity formulation variant TN-ISM (center), and open and interconnected porosity formulation variant TN-ADBS (right). Scanned using Scanco μCT 40 (Scanco, Brüttisellen, Switzerland; 70kVp 34μAs) and analyzed using Python (v3, Pydicom module).

Formulation	Porosity (%)	Maximum torque at $t = 0$ (Ncm)	Maximum torque at $t = 6$ weeks (Ncm)
TN-SM	2,67	32,4 ± 4,8	18,1 ± 5,1
TN-ISM	29,9	17,1 ± 2,2	23,1 ± 7,2

Table 4.
Comparison of reverse torque to failure values for dense and porous formulation variants TN-SM and TN-ISM in vivo, *at* t = 0 *and at 6 weeks post-placement using Ø3,7x8mm Tapered Screw Vent Implant (Zimmer, Warsaw, Indiana).*

torque required with the porous formulation was reduced by less than half when evaluated in artificial bone substrates at T = 0. Furthermore, it is interesting to note that the initially lower strength of the more porous TN-ISM rises rapidly, resulting in higher value for the porous TN-ISM than dense TN-SM at 6 weeks post-implantation *in vivo*. This behavior is likely the result of the material being more rapidly penetrated and replaced by osseointegrating bone than the dense formulation. The level of stabilization afforded by both formulations of the adhesive, even with marginally short and thin implants, Ø3,7 x 8 mm, is clinically acceptable.

It should be noted that different clinical applications may be better served by longer, or shorter, duration of transition from the biomaterial to bone depending on expected level of stress, shock load, expected material fatigue, need to serve as an implant bed, and other clinical, mechanical or biological considerations. The pursuit of levers of control over such functional parameters as compressive or tensile strength, or the rate of degradation and bone substitution, remains an active driver of this adhesive biomaterial development. Certain clinical applications, e.g., dental implant stabilization, present very different requirements than a grafting material for cystic cavities, spinal fusion, long bone fracture fixation or cranial flap fixation. In addition, the control allowed by alteration of the physical state of the material provides for modifications at the much smaller chemical scale, discussed in the following section.

5.3 Enhanced strength with slower degradation and bone substitution

The addition of citrate species, $H_xC_6H_5O_7^{3-x}$ ($0 \leq x \leq 3$), which readily substitute for OPLS in the formation of a solid mass, results in profound changes in both biomechanical and biological response. For example, when combinations of tribasic calcium citrate (CaCit) and citric acid (HCit) were added through partial substitution of OPLS, the resulting reaction mixture is significantly more tacky, more cohesive, and results in a solid which is both physically stronger (**Figure 18**) and more resistant to *in vivo* degradation (**Figure 19**).

The biological response to the citrate-modified formulation variant is presented in **Figure 19**. As evident in the photomicrographs, the dense TN-SM deposit (top) has undergone a significant amount of bone substitution as well as subperiosteal bone deposition. In contrast, the citrate containing formulation (bottom) has been neither significantly penetrated nor coated by new bone tissue during the 16-week

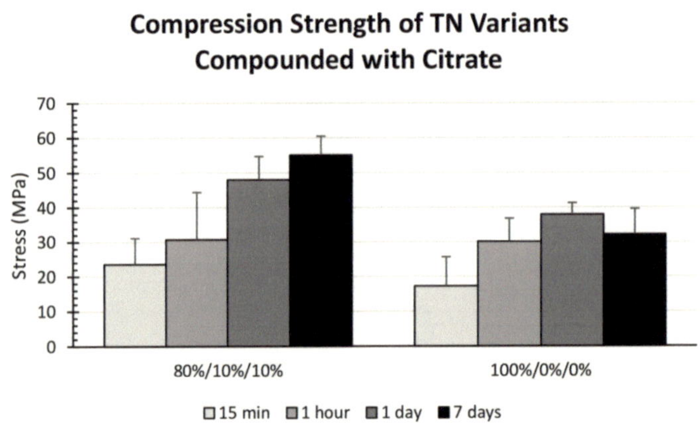

Figure 18.
Comparison of time dependence of compression strength of TN-SM with a formulation variant containing citrate, 80% OPLS, 10% $C_6H_3O_7$ *(citric acid) and 10%* $Ca_3\ (C_6H_5O_7)_2$ *(calcium citrate). Note the general trend of increasing strength of both formulations with time and the increase in compressive strength associated with addition of citrate.* (n = 3).

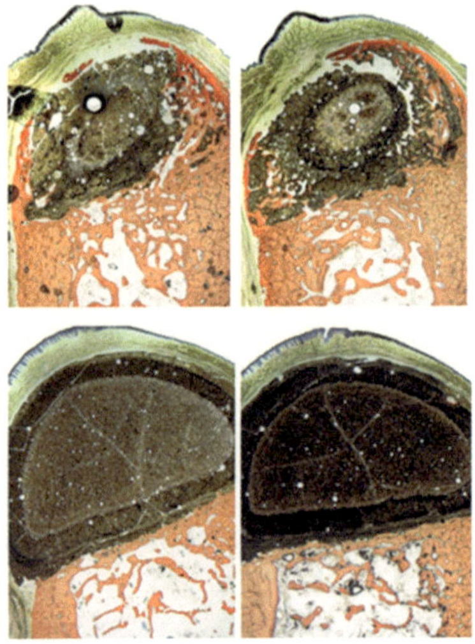

Figure 19.
Photomicrographs of transverse section through experimental box defects in mandibular residual ridge restored 16 weeks previous with either dense Tetranite adhesive TN-SM (top) or citrate containing variant 80/10/10 (bottom) (canine mandibular residual ridge).

in vivo residence. The model used healed canine residual ridges into which 7 mm H x 8 mm L, through-and-through defects were prepared immediately before the implantations [56]. No membranes were used with either adhesive formulation.

5.4 Vehicle for administration of a therapeutic - antimicrobial

The use of bone cements and regenerative materials is often associated with injury, such as a bone fracture, and always with a surgical intervention. Often the host site is frankly contaminated or at risk of being contaminated, and the procedures are associated with a significant risk of serious infection to follow [57]; therefore, the inclusion of locally active drugs to prevent or treat these in the formulation could be highly beneficial. Hence, it is not surprising that extensive effort has been directed for over 150 years [58] to compounding various bone cements with various antimicrobials such as antibiotics [59–62] additional metal ions, [63] and antimicrobial nanometals [61, 64]. It should be noted that bone cements, even calcium phosphate-based bone cements, are not necessarily bone adhesives. The potential use of calcium phosphate-OPLS based bone adhesives such as the ones discussed in this chapter, likewise, is being investigated. A recent study shows promise in formulations which include a combination of two antibiotics, vancomycin and amikacin, in a contaminated deep sternal wound animal model using a mineral-organic adhesive [22].

5.5 Vehicle for administration of a therapeutic: Pain control

Trauma and surgery are often associated with need for pain control, and incorporation of local anesthetic for local release from a calcium phosphate bone cement has been described in the literature [65]. The authors reported a burst release of the anesthetic the rate of which depended on the cement pH. The adhesives described in this chapter can also serve as a vehicle for timed release of pain control agents. An example of a clinical situation in which this ability may be used clinically is during recovery from extraction of wisdom teeth. It is standard practice to prescribe opioids for patients undergoing third molar odontectomies. The opioid drugs often prescribed

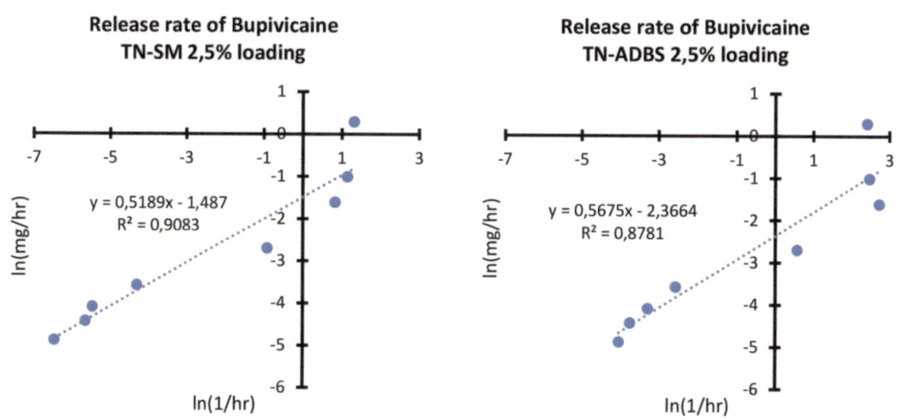

Figure 20.
Log-log linearity in the release kinetics of bupivacaine into simulated tissue fluid medium from a dense formulation TN-SM and a porous formulation TN-ISM of Tetranite bone regenerative adhesive. Data represent 1 week of release kinetics.

may lead to dependence and misuse issues. Administration of a locally acting anesthetic with appropriate release kinetics can provide an effective alternative pain control strategy with much-diminished risks. Bupivacaine has presented multiple advantages over other local anesthetics, for example, lidocaine, articaine, etc., as well as other classes of potential locally acting pain control drugs, for example, NSAIDs, gabapentinoids, etc. Among these are chemical stability in the environment, stability to sterilization procedures, favorable release kinetics, and good tissue retention. **Figure 20** illustrates the release kinetics *in vitro* of bupivacaine from two formulations of Tetranite adhesive, one porous and one dense, over a time scale of approximately 1 week.) Preclinical trial of safety and efficacy of anesthetic releasing formulation of the bone regenerative adhesive is currently pending in a canine odontectomy model.

6. Discussion and conclusions

Tetranite adhesives are a promising novel class of mineral-organic bioactive materials which offer unique combinations of properties. As demonstrated in this chapter, these include robust wet-field adhesion generating strong and durable bonds *in vivo*, excellent biocompatibility, osteoconductivity combined with bioactivity and bone growth stimulative effect leading to total replacement by bone within 3 months to a year, injectability to the site of action, self-setting within minutes to a weight-bearing strength, and flexibility in formulation allowing modification of critical attributes. This flexibility includes the ability to modify the handling properties, mechanical properties, therapeutic elution kinetics, and resorption kinetics, as well as other biological response parameters depending on the clinical requirements, by design. Although the exemplary formulations based on the platform adhesive described and discussed in this chapter touch on several of these clinically important characteristics and potential applications, other potential applications, such as the use of dense adhesive formulation for fracture fixation, the use its highly porous formulation as a bone void filler replaced by osseous tissues within a few weeks, and the use of the material as a drug carrier are currently being investigated through a number of preclinical as well as clinical studies. Through attainment of their unique combination of attributes, the calcium organo-phosphate adhesives offer the potential to fulfill a set of clinical requirements and needs that has so far not been satisfied by any of the biomaterials available on the market, including calcium phosphate cements, polyurethane and polymethylmethacrylate bone cements, particulate bone void fillers, and others.

Acknowledgements

This research was supported by *the International Team for Implantology (ITI)*, *National Institute of Neurological Disorders and Stroke (NINDS)*, *National Institute of Dental and Craniofacial Research (NIDCR)*, and *National Institute of Aging (NIA)* of the National Institutes of Health under award numbers 1277_2017, R44NS115386, R43DE029369, R43AG079741, R44DE029369, and R44DE032564.

All data and figures were internally generated unless cited, and none of the material has been previously published unless specifically referenced.

Many thanks to Hari Prasad, Hard Tissue Research Laboratory, University of Minnesota for expert histological preparations.

All animal use has been approved by the applicable Institutional Animal Care and Use Committees.

Conflict of interest statement

GK and MB are co-founders of RevBio, hold stock and stock options and are currently employed by the company. BM, SB, and RJ are employees of RevBio and hold stock options.

Author details

George W. Kay*, Rahul Jadia, Sourabh Boruah, Brittany McDonough and Michael C. Brown
RevBio, Inc., Lowell, MA, USA

*Address all correspondence to: gkay@revbio.com

References

[1] Wang CS, Stewart RJ. Multipart copolyelectrolyte adhesive of the sandcastle worm, *Phragmatopoma californica* (Fewkes): Catechol oxidase catalyzed curing through peptidyl-DOPA. Biomacromolecules. 2013;**14**(5):1607-1617

[2] Zhao H, Sun C, Stewart RJ, Waite JH. Cement proteins of the tube-building polychaete *Phragmatopoma californica*. The Journal of Biological Chemistry. 2005;**280**(52):42938-42944

[3] Li Q, Tang B, Liu X, Chen B, Wang X, Xiao H, et al. Overcoming the dilemma of *in vivo* stable adhesion and sustained degradation by the molecular design of polyurethane adhesives for bone fracture repair. Advanced Healthcare Materials. 2024;**13**(5):2301870

[4] Schreader KJ, Bayer IS, Milner DJ, Loth E, Jasiuk I. A polyurethane-based nanocomposite biocompatible bone adhesive. Journal of Applied Polymer Science. 2013;**127**(6):4974-4982

[5] Vieira JS, Santos FR, de Freitas JV, Baratto-Filho F, Gonzaga CC, de Araujo MR. Bond strength evaluation of cyanoacrylate-based adhesives and screws for bone fixation. Oral and Maxillofacial Surgery. 2016;**20**(2):157-160

[6] Kim B-S, Sung H-M, You H-K, Lee J. Effects of fibrinogen concentration on fibrin glue and bone powder scaffolds in bone regeneration. Journal of Bioscience and Bioengineering. 2014;**118**(4):469-475

[7] Morilla, Espino C, Estévez GF, van der Weerd L, de Geus-Oei L-F, van den Beucken JJJP. Innovations for brushite cements toward applications in bone regeneration and drug delivery. Ceramics International. 2024

[8] Kesseli FP, Lauer CS, Baker I, Mirica KA, Van Citters DW. Identification of a calcium phosphoserine coordination network in an adhesive organo-apatitic bone cement system. Acta Biomaterialia. 2020;**105**:280-289

[9] Kanehisa Laboratories. Glycine, Serine and Threonine Metabolism - Homo sapiens (Human). Kyoto, Japan: Kanehisa Laboratories; 2023. Available from: https://www.genome.jp/kegg-bin/show_pathway?hsa00260+C01005

[10] Pawson T, Scott JD. Protein phosphorylation in signaling – 50 years and counting. Trends in Biochemical Sciences. 2005;**30**(6):286-290

[11] Beck GR, Zerler B, Moran E. Phosphate is a specific signal for induction of osteopontin gene expression. National Academy of Sciences of the United States of America. 2000;**97**(15):8352-8357

[12] Park J-W, Kim Y-J, Jang J-H, An C-H. MC3T3-E1 cell differentiation and *in vivo* bone formation induced by phosphoserine. Biotechnology Letters. 2011;**33**(7):1473-1480

[13] Farrar DF. Bone adhesives for trauma surgery: A review of challenges and developments. International Journal of Adhesion and Adhesives. 2012;**33**:89-97

[14] Cui C, Liu W. Recent advances in wet adhesives: Adhesion mechanism, design principle and applications. Progress in Polymer Science. 2021;**116**:101388

[15] Norton MR, Kay GW, Brown MC, Cochran DL. Bone glue - the final

frontier for fracture repair and implantable device stabilization. International Journal of Adhesion and Adhesives. 2020;**102**:102647

[16] Panagiotopoulou VC, Santolini E, Jones E, Jha A, Giannoudis PV. Adhesives for treatment of bone fractures: A review of the state-of-the art. Injury. 2022;**53**:S20-SS5

[17] Shokri M, Dalili F, Kharaziha M, Baghaban Eslaminejad M, Ahmadi TH. Strong and bioactive bioinspired biomaterials, next generation of bone adhesives. Advances in Colloid and Interface Science. 2022;**305**:102706

[18] Bingol HB, Bender JCME, Opsteen JA, Leeuwenburgh SCG. Bone adhesive materials: From bench to bedside. Materials Today Bio. 2023;**19**:100599

[19] Kirillova A, Nillissen O, Liu S, Kelly C, Gall K. Reinforcement and fatigue of a bioinspired mineral-organic bioresorbable bone adhesive. Advanced Healthcare Materials. 2021;**10**(2):e2001058

[20] Cochran DL, Jones AA, Sugita R, Brown MC, Prasad H, Kay GW. Twelve-month evaluation of a novel mineral-organic adhesive material used to stabilize dental implants placed in oversized osteotomies *in vivo* in an animal model. Clinical Oral Implants Research. 2022;**33**(4):391-404

[21] Foley KT, Woodard EJ, Slotkin JR, Mayotte CK, Baldwin AC, Brown MC, et al. Cranial flap fixation in sheep using a resorbable bone adhesive. Journal of Neurosurgery. 2020;**134**:1-9

[22] Pradeep A, Varma PK, Arumugam T, Vijayan AM, Vasudevan AK, Rangasamy J. Antibacterial bone adhesive cement for preventing sternal infections after cardiac surgery. Ceramics International. 2023;**49**(10):16110-16122

[23] Jadia R, Bichara DA, Wheelis SE, Allen G, Hess B, Kay GW. A novel opioid-free targeted pain control method for acute post-operative localized pain related to Oral surgical procedures. In: NIH HEAL Initiative Annual Meeting. 2022

[24] Kay GW, Brown M, Pikos M, Pollack A, Fugazzotto P, Donoff B, et al. Next-generation osteoconductive resorbable bone adhesives: Tetranite. In: Miron RJ, Zhang, Yufeng, editor. Next-Generation Biomaterials for Bone & Periodontal Regeneration. Batavia, IL: Quintessence Publishing Co, Inc.; 2019. p. xiv, 370 pages

[25] Cochran DL, Jones A, Sugita R, Brown MC, Guda T, Prasad H, et al. Immediate dental implant stabilization in a canine model using a novel mineral-organic adhesive: 4-month results. The International Journal of Oral & Maxillofacial Implants. 2020;**35**(1):39-51

[26] Kirillova A, Kelly C, von Windheim N, Gall K. Bioinspired mineral-organic bioresorbable bone adhesive. Advanced Healthcare Materials. 2018;**7**(17):e1800467

[27] Sugita R, Jones AA, Kotsakis GA, Cochran DL. Radiographic evaluation of a novel bone adhesive for maintenance of crestal bone around implants in canine oversized osteotomies. Journal of Periodontology. 2022;**93**(6):924-932

[28] Suga T, Okabe N. Aqua(l-O-serine phosphato) calcium(II). Acta Crystallographica Section C. 1996;**52**(8):1894-1896

[29] Śmiechowski M. Theoretical pKa prediction of O-phosphoserine in

aqueous solution. Chemical Physics Letters. 2010;**501**(1):123-129

[30] USFDA. Use of International Standard ISO 10993-1: "Biological Evaluation of Medical Devices - Part 1: Evaluation and Testing within a Risk Management Process": Guidance for Industry and Food and Drug Administration Staff. Silver Spring, Maryland, USA: USFDA; 2023

[31] Lin FH, Liao CJ, Chen KS, Su JS, Lin CP. Petal-like apatite formed on the surface of tricalcium phosphate ceramic after soaking in distilled water. Biomaterials. 2001;**22**(22):2981-2992

[32] Wang B, Zhang Z, Pan H. Bone apatite nanocrystal: Crystalline structure, chemical composition, and architecture. Biomimetics [Internet]. 2023;**8**(1):90. DOI: 10.3390/biomimetics8010090

[33] Sugita R, Jones AA, Kotsakis GA, Cochran DL. Radiographic evaluation of a novel bone adhesive for maintenance of crestal bone around implants in canine oversized osteotomies. Journal of Periodontology. 2021:1-9. DOI: 10.1002/JPER.20-0876

[34] Pikos MA, Miron RJ. Clinical use of a mineral-organic resorbable bone adhesive: A first human case report with a 3-year follow-up. The International Journal of Periodontics & Restorative Dentistry. 2024;**44**(1):51-57

[35] Smith T, Foley K, Boruah S, Slotkin J, Woodard E, Lazor JB, et al. Use of adhesive cranial bone flap fixation without hardware to improve mechanical strength, resist cerebrospinal fluid leakage, and maintain anatomical alignment: A laboratory study. Journal of Neurosurgery. 16 Dec 2022;**139**(2):517-527. DOI: 10.3171/2022.10.JNS221657 Print 2023 Aug 1

[36] Salinas AJ, Vallet-Regí M. Bioactive ceramics: From bone grafts to tissue engineering. RSC Advances. 2013;**3**(28):11116-11131

[37] Boden SD, Schimandle JH, Hutton WC. An experimental lumbar intertransverse process spinal fusion model. Radiographic, histologic, and biomechanical healing characteristics. Spine (Phila Pa 1976). 1995;**20**(4):412-420

[38] Meredith N. Assessment of implant stability as a prognostic determinant. The International Journal of Prosthodontics. 1998;**11**(5):491-501

[39] Norton MR. First-in-human pilot study to assess methodology for using a mineral-organic bone adhesive for optimization of primary stability and implant success for implants glued into immediate extraction sockets and immediately temporized: Pilot study. The International Journal of Oral & Maxillofacial Implants. 2024;**39**(1):29-39

[40] Wu J, Liu F, Wang Z, Liu Y, Zhao X, Fang C, et al. The development of a magnesium-releasing and long-term mechanically stable calcium phosphate bone cement possessing osteogenic and immunomodulation effects for promoting bone fracture regeneration. Frontiers in Bioengineering and Biotechnology. 2021;**9**:803723

[41] Demirel M, Kaya AI. Effect of strontium-containing compounds on bone grafts. Journal of Materials Science. 2020;**55**(15):6305-6329

[42] Vestermark MT. Strontium in the bone-implant interface. Danish Medical Bulletin. 2011;**58**(5):B4286

[43] Ding S-J, Wei C-K, Lai M-H. Bio-inspired calcium silicate–gelatin bone grafts for load-bearing applications.

Journal of Materials Chemistry. 2011;**21**(34):12793-12802

[44] Sanmartin de Almeida M, Fernandes GVO, de Oliveira AM, Granjeiro JM. Calcium silicate as a graft material for bone fractures: A systematic review. The Journal of International Medical Research. 2018;**46**(7):2537-2548

[45] Ruiz-Agudo E, Ruiz-Agudo C, Di Lorenzo F, Alvarez-Lloret P, Ibanez-Velasco A, Rodriguez-Navarro C. Citrate stabilizes hydroxylapatite precursors: Implications for bone mineralization. ACS Biomaterials Science & Engineering. 2021;7(6):2346-2357

[46] Costello LC, Chellaiah M, Zou J, Franklin RB, Reynolds MA. The status of citrate in the hydroxyapatite/collagen complex of bone; and its role in bone formation. Journal of Tissue Engineering and Regenerative Medicine. 2014;**3**:4

[47] Koch M, Appoloni O, Haufroid V, Vincent JL, Lheureux P. Acute barium intoxication and hemodiafiltration. Journal of Toxicology. Clinical Toxicology. 2003;**41**(4):363-367

[48] Barralet J, Best S, Bonfield W. Carbonate substitution in precipitated hydroxyapatite: An investigation into the effects of reaction temperature and bicarbonate ion concentration. Journal of Biomedical Materials Research. 1998;**41**(1):79-86

[49] Sheikh Z, Abdallah MN, Hanafi AA, Misbahuddin S, Rashid H, Glogauer M. Mechanisms of *in vivo* degradation and resorption of calcium phosphate based biomaterials. Materials (Basel). 2015;**8**(11):7913-7925

[50] Vainionpaa S, Vihtonen K, Mero M, Patiala H, Rokkanen P, Kilpikari J, et al. Fixation of experimental osteotomies of the distal femur of rabbits with biodegradable material. Archives of Orthopaedic and Trauma Surgery (1978). 1986;**106**(1):1-4

[51] Jiao J, Hong Q, Zhang D, Wang M, Tang H, Yang J, et al. Influence of porosity on osteogenesis, bone growth and osteointegration in trabecular tantalum scaffolds fabricated by additive manufacturing. Frontiers in Bioengineering and Biotechnology. 2023;**11**:1117954

[52] Karageorgiou V, Kaplan D. Porosity of 3D biomaterial scaffolds and osteogenesis. Biomaterials. 2005;**26**(27):5474-5491

[53] Luo C, Wang C, Wu X, Xie X, Wang C, Zhao C, et al. Influence of porous tantalum scaffold pore size on osteogenesis and osteointegration: A comprehensive study based on 3D-printing technology. Materials Science & Engineering. C, Materials for Biological Applications. 2021;**129**:112382

[54] Phadke A, Hwang Y, Kim SH, Kim SH, Yamaguchi T, Masuda K, et al. Effect of scaffold microarchitecture on osteogenic differentiation of human mesenchymal stem cells. European Cells & Materials. 2013;**25**:114-129

[55] Wang P, Zhao L, Liu J, Weir MD, Zhou X, Xu HH. Bone tissue engineering via nanostructured calcium phosphate biomaterials and stem cells. Bone Research. 2014;**2**:14017

[56] Schenk RK, Buser D, Hardwick WR, Dahlin C. Healing pattern of bone regeneration in membrane-protected defects: A histologic study in the canine mandible. The International Journal of Oral & Maxillofacial Implants. 1994;**9**(1):13-29

[57] Borger MAMD, Rao VMD, Weisel RDMD, Ivanov JM, Cohen GMD,

Scully HEMD, et al. Deep sternal wound infection: Risk factors and outcomes. The Annals of Thoracic Surgery. 1998;**65**(4):1050-1056

[58] Kuris EO, Osorio C, Anderson GM, Younghein JA, McDonald CL, Daniels AH. Utilization of antibiotic bone cement in spine surgery: Pearls, techniques, and case review. Orthopedic Reviews (Pavia). 2023;**15**:90618

[59] Ratier A, Freche M, Lacout JL, Rodriguez F. Behaviour of an injectable calcium phosphate cement with added tetracycline. International Journal of Pharmaceutics. 2004;**274**(1):261-268

[60] Alkhraisat MH, Rueda C, Cabrejos-Azama J, Lucas-Aparicio J, Mariño FT, Torres García-Denche J, et al. Loading and release of doxycycline hyclate from strontium-substituted calcium phosphate cement. Acta Biomaterialia. 2010;**6**(4):1522-1528

[61] Lin H, Gao Z, Shan T, Asilebieke A, Guo R, Kan YC, et al. A review on the promising antibacterial agents in bone cement-from past to current insights. Journal of Orthopaedic Surgery and Research. 2024;**19**(1):673. DOI: 10.1186/s13018-024-05143-7

[62] Ginebra MP, Traykova T, Planell JA. Calcium phosphate cements as bone drug delivery systems: A review. Journal of Controlled Release. 2006;**113**(2):102-110

[63] Jayasree R, Sampath Kumar TS, Perumal G, Doble M. Drug and ion releasing tetracalcium phosphate based dual action cement for regenerative treatment of infected bone defects. Ceramics International. 2018;**44**(8):9227-9235

[64] Wekwejt M, Michno A, Truchan K, Pałubicka A, Świeczko-Żurek B, Osyczka AM, et al. Antibacterial activity and cytocompatibility of bone cement enriched with antibiotic, nanosilver, and nanocopper for bone regeneration. Nanomaterials (Basel). 2019;**9**(8):1114. DOI: 10.3390/nano9081114

[65] Irbe Z, Loca D, Vempere D, Berzina-Cimdina L. Controlled release of local anesthetic from calcium phosphate bone cements. Materials Science and Engineering: C. 2012;**32**(6):1690-1694